TRIBAL SOLDIERS OF VIETNAM

TRIBAL SOLDIERS OF VIETNAM

The Effects of Unconventional Warfare on Tribal Populations

David K. Moore

ISBN 10:	Hardcover	1-4257-4474-5
	Softcover	1-4257-4473-7
ISBN 13:	Hardcover	978-1-4257-4474-8
	Softcover	978-1-4257-4473-1

This book was printed in the United States of America.

To order additional copies of this book, contact:
Xlibris Corporation
1-888-795-4274
www.Xlibris.com
Orders@Xlibris.com

37679

CONTENTS

Introduction 11

I: Communist Insurgency Warfare 23
Confucianism and Communism: A Blueprint for Victory 23
Parallel Hierarchies: Ruling Through the Masses 28
Communist Revolutionary Warfare: Warfare Through the Masses 30
Propaganda and the Secret Police: Controlling the Masses 32

II: Counterinsurgency Warfare 35
Malaya and the Philippines: The Basis for American Theory 35
Trinquier and Modern Warfare: Tribal Manipulation 38
Operation X: Into the Milieu 42
The Buon Enao Experiment: Winning the "Hearts and Minds" 44
Revolutionary Development and CAP's: Fighting Fire With Fire 46
Phoenix and the CTT's: Losing the Hearts and Minds 48
RW::(PW plus GW) 49
COIN::GW 50

III: Case Studies From North Vietnam 51
The Tho 52
The Tho and the Vietnam People's Army 53
The Battle of R.C. 4: End of a Colonial Era 54
The Tai Highlands 56
World War II and After: the Militarization of the Tai 58

Why Defend the Tai Highlands at All? 60
The Evacuation of Lai Chau: Defeat of a Tribal Army 61
Dien Bien Phu: A Clash of Ethnic Armies 64

IV: Case Studies From South Vietnam 65
Mayrena and the Catholic Mission in Kontum: A Heart of Darkness? 66
Pacification: Consolidating the Highland Minorities 67
The First Indochina War 70
Second Indochina War: Tribal Mobilization (Again) 72
CIDG After 1963: The Conventionalization of Tribal Troops (Again) 74
The 1964 Revolt in the Highlands 75
Summary of Chapter III and IV 77

V: An Anthropological Interpretation of Change 78
North Vietnam After the First Indochina War 79
Parallel Hierarchies: the Assimilation of Tribal Leadership 81
Front Organizations and Land Reform: Assimilation of the Masses 82
Leadership and Western Counterinsurgency in South Vietnam 84
Crime and Counterinsurgency: The Abuse and Corruption of Power 86
Ethnonationalist Movements: A Counterinsurgency Byproduct? 91
Summary of Findings: Barth and the Unconventional Warfare Formulae 95

Bibliography 99

Endnotes 103

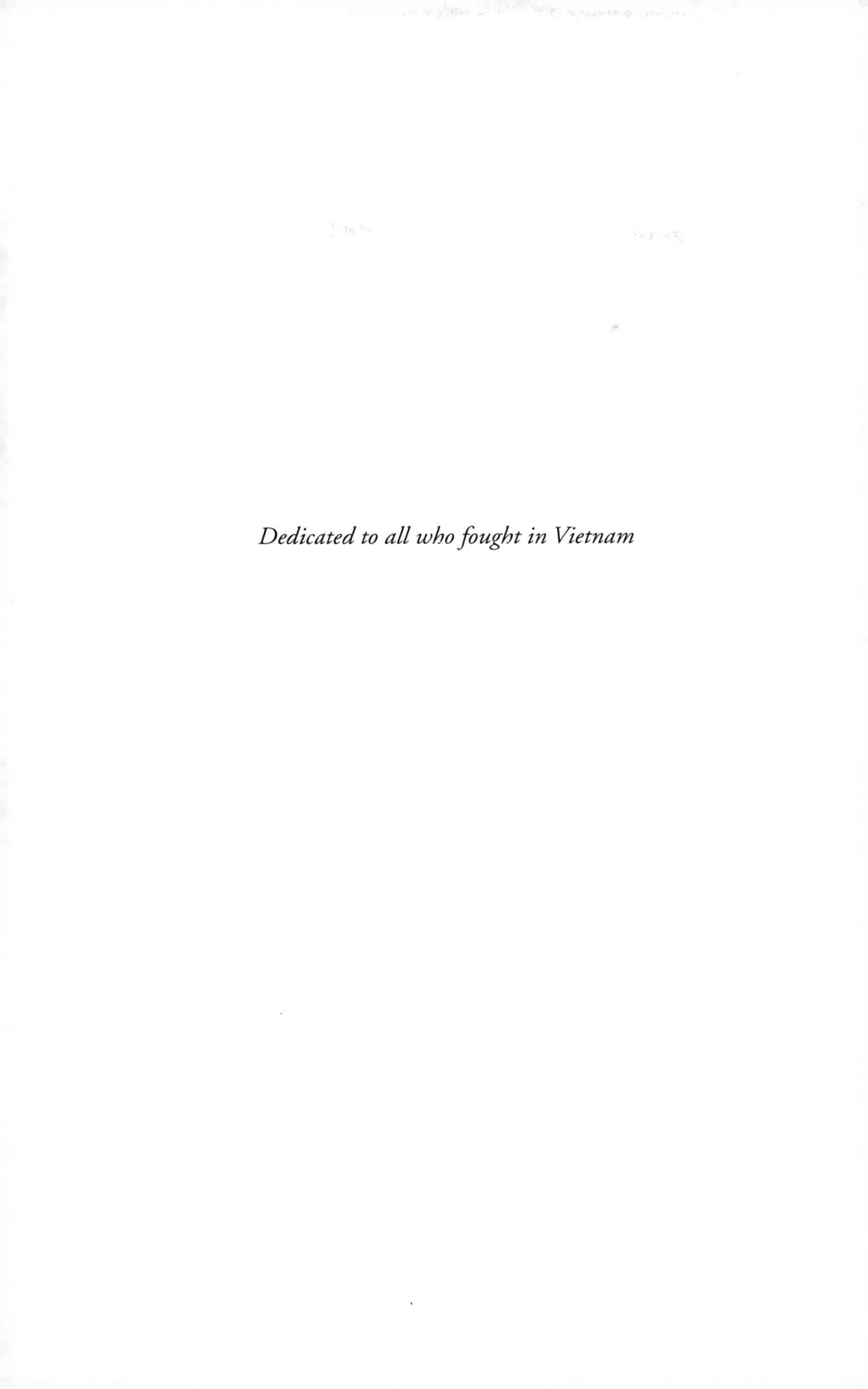

Dedicated to all who fought in Vietnam

PREFACE

I INITIALLY WROTE THIS thesis while a graduate student at California State University, Sacramento, from 1982-84. Although I wrote this study over twenty years ago, I feel it is just as valuable today as then. The problems with arming tribal factions in Afghanistan as well as militias in Iraq are all too evident when viewed through the prism of tribal soldiers in Vietnam. I decided not to update any information from my initial thesis. Aside from the fact that this would require more work on my part, I found it unnecessary from the standpoint of elucidating the dangers of tribal arming for modern political purposes.

There is also a story connected with this thesis. This thesis was authorized, sanctioned, and initially assisted by a variety of professors in the Anthropology Department at CSUS. A few of them voiced "concern" that the findings could be "misused by the CIA," whatever that meant. As an example of the sorry state of academia in America, it never occurred to these professors that they had an ethical obligation to complete a task, especially since I had completed all classes with a perfect 4.0. I had moved to Washington, DC, and from there submitted my first complete draft, which is what the reader sees.

My thesis committee immediately resigned and now stated the thesis was "improper" and that I had to not only constitute a new committee with a new subject, but had to complete all new tasks within eighteen months in order to obtain my MA. This was an impossibility and they knew it. This was simply a group of Leftist professors imposing their political beliefs, in that I was now subverting the "ethics" of Anthropology for some nefarious purpose in the corridors of power in Washington. How

I was supposedly manipulating the world while working as a restaurant manager in Alexandria, Virginia, did not seem illogical to them.

My only choice in trying to obtain the MA I worked so hard for was to convene a student trial. But, as one of the professors involved in the trial stated, "I have to work here and you are in DC." I personally liked the professor, but this is why statues are erected for those who tell the truth. It always is a tough decision made, unfortunately, by the few. I even appealed directly to the president of CSUS when he came to Washington. I knew he would never call his professors on the carpet for this type of highly unethical behavior, but at least I wanted to prove that he was just as spineless as the rest of the university. He was.

With the new technology of the Internet and print-on-demand, however, studies such as mine can never be suppressed. The social, political, and military ramifications of the arming of tribesmen is something all Americans should be aware of. When the American government funnels a billion dollars to tribal groups, in secret or otherwise, with no discussion in Congress or the general public of the ultimate consequences, somebody isn't doing their job. How Al Qaeda evolved out of our arming of tribal groups to fight the Soviets is not a surprise to me or anybody who takes the time to read and study my book. If that constitutes "misuse" of Anthropology then consider me guilty.

In the final analysis, I don't feel all that bad at being denied my MA from CSUS, especially when you look at the sorry professors who did the denying. I also have good company, as fellow combat veteran Kurt Vonnegut was initially denied his MA in Anthropology at the University of Chicago after submitting his anthropology thesis, his thesis committee considering Vonnegut's thesis "unprofessional." Interestingly, Vonnegut's thesis was also related to unconventional warfare. His thesis concerned the relation between Cubist art and Native American uprisings of the 19th Century. At least that university many years later saw their error and awarded Vonnegut his proper MA. I can only hope that the California State University, Sacramento, will eventually right their own wrong sometime in the future regarding my MA. Since I work at the Library of Congress as a German Acquisitions Specialist I am easy to track down.

INTRODUCTION

THIS THESIS IS a brief examination and analysis of specific affects of unconventional warfare practices on affected tribal populations. The vast majority of wars since World War II have been "unconventional" in nature, i.e., guerrilla warfare, border warfare, revolutionary warfare, etc. As a rule, unconventional forms of warfare can be pinpointed to areas that are rural and technologically underdeveloped, with the Third World today being the arena for unconventional warfare. Such countries are poor and unable to afford the expensive equipment that "conventional" armies possess, such as tanks, artillery, and aircraft. Poor countries are therefore forced to resort to more economical or "unconventional" methods of warfare such as guerrilla tactics (lightly armed mobile infantry), sabotage, or psychological operations (i.e., propaganda, terror, etc.).

This author's own interest in the subject of unconventional warfare began early with the reading of such classics as *The Seven Pillars of Wisdom* by T. E. Lawrence, more popularly known as Lawrence of Arabia. While serving in conventional airmobile and airborne units in Vietnam and the U.S., this author became acquainted with unconventional warfare practices while in the U.S. Army. This interest was to continue from military life into the realm of academia. While studying Near Eastern Archaeology as a graduate student at the Freie Universitaet Berlin, some emphasis was placed by the department on various tribal groups that had been contracted for military service by city-states such as Elam, Ur, and Babylonia. While working in the Negev Desert as a surveyor for the Israeli

Department of Antiquities this author came in contact with Bedouins who had served either in the "Turkish War," what the Bedouins call World War I, or with the Israeli Defense Forces. The cumulative effect of such academic studies and personal experiences has naturally led to a thesis on tribal nation-state military service and the effects of such service on the tribal population.

Although this thesis relies heavily on historical data, it should be of interest primarily to the anthropologist. This is not simply because the thesis concerns tribal populations, the primary domain of anthropological studies, but also because anthropology in many instances is now analyzing how various cultures change through time. Many such anthropological articles and books concerning culture change deal with such areas as economic modernization or trade. Anthropologists such as Lee and Hurlich (see below) have written one study concerning South Africa's recruitment of Namibian San Bushmen as unconventional soldiers, but such warfare studies are small in comparison.

The question is why anthropologists should devote more attention to warfare studies. This author's answer is that modern warfare, particularly any brand of warfare that dictates the arming of tribal populations, should be closely observed by anthropologists. There should be no reason to doubt that modern militarization has and will continue just as profoundly to disrupt tribal populations as for example urbanization or industrialization. Although any number of indigenous institutions as pertains to local trade or economy could have been targeted for a warfare study, the primary emphasis of this thesis has been directed to tribal leadership.

Interestingly, a number of "Western" leadership patterns were to undergo change as a direct consequence of the First and Second World Wars. In World War I, due to the nature of trench warfare and the military theory of "attrition," huge numbers of men were dying on the battlefield. The character of the British Army underwent such a dramatic change that it was eventually referred to as two distinct armies: the "old army" and the "new army." In the old army, traditional leadership was restricted to the aristocrats. As attrition set in, the officers' ranks were no longer restricted to a social elite. Any man who could soldier well could be promoted from the enlisted ranks up to even lieutenant colonel (Keegan, 1976:272). This is not to say that the new officer corps was readily accepted by the old, as the former was referred to as "temporary gentlemen" by the latter.

During World War II, the leadership pattern of the British Army would again undergo change through the initiation of the written examination. Officers were to be selected on the basis of " . . . intelligence, stability, companionability, leadership potential and the like, conditions which favored the middle-class over the working-class" (Keegan, 1976:272). The U.S. Army also used examinations for Officer Candidate School (OCS), an examination with which the author is personally familiar. During the latter part of the Vietnam War the examination method was still in effect, but the upper middle-class was being further favored by the prerequisite of at least two years of college education for officer candidacy.

Although the above-mentioned examples concern the nation-state population engaged in conventional warfare, are there any similar patterns of leadership change concerning a tribal population engaged in unconventional warfare? When a tribal population is targeted for recruitment, what factors are taken into consideration by the nation-state determining who the tribal military leaders will be? Just as there was an "old" and "new" military elite (i.e. officer class) in the British Army, are there "old' and "new" elites in tribal societies due to a particular style of warfare?

The primary anthropological framework chosen by this author to analyze such a possible tribal political shift is Barth's New Elite Theory. This theory was not developed by Barth to study unconventional forms of warfare, but was adapted by the author for this purpose. Central to Barth's theory is the role of individuals from an ethnic group who come in contact with advanced societies. These individuals are classified as "cultural brokers" and in Barth's opinion come to constitute "agents of change." These "agents of change" will eventually become members of a "new elite" among their ethnic population. Although it is not specifically stated by Barth, it is assumed by this author that it is irrelevant whether or not the cultural brokers serve a military or economic function.

Once the cultural brokers are in place in the tribal society, the society in general becomes aware of a number of options for their future behavior in relation to the more advance society. They can:

1) Become incorporated into the larger society. Result: the ethnic groups would probably remain at the low end of the socio-economic scale and remain culturally conservative.
2) Accept minority status in their own cultural areas and reduce their minority disabilities through education, for example. Result: the groups would probably become assimilated.
3) Emphasize ethnic identity, which would be used to " . . . develop new positions and patterns to organize activities in those sectors formerly not found in their society." Result: creates varied movements from nativism to new states.

Provided that Barth's third option applies to an affected tribal group, Geertz's theory of integration would make for an interesting correlation. Whereas Barth's third option concerns a more pronounced ethnic identity and a desire for separation, Geertz's theory explores the relationship between what he terms the "primordial ties" of the ethnic group and the "nation" itself.

Geertz has listed six primordial ties, or sentiments, that can bind an ethnic group together: assumed blood ties, race, language, region, religion, and custom. It is the first that is viewed by Geertz of primary importance to Southeast Asia, even though Geertz informs us that more than one sentiment may be present. Geertz defines the first as:

> **Assumed Blood Ties**. Here the defining element is Quasikinship. 'Quasi' because kin units formed around known biological relationships (extended families, lineages, and so on) are too small for even the most tradition-bound to regard them as having more than limited significance, and the referent is, consequently, to a nation of untraceable but yet sociologically real kinship, as in a tribe. Nigeria, the Congo, and the greater part of sub-Saharan Africa are characterized by a prominence of this sort of primordialism. But so also

> are the nomads or seminomads of the Middle East—the Kurds, Baluchis, Pathans, and so on; the Nagas, Mundas, Santals, and so on, of India; *and most of the so-called 'hill tribes' of Southeast Asia.* (Geertz, 0000:112, italics added).

As with Barth, Geertz's theory was not originally developed as a means to study unconventional warfare, but was adapted this author. Of particular interest should be the role of the military as a process of integration for tribal minorities. One obvious question that will hopefully be answered by this thesis is whether or not primordial ties, specifically assumed blood ties, conflict with the social changes brought about by the nation-state's politico-military hierarchy being imposed on the tribal group.

A number of recent authors have either superficially examined the problem of tribal militarization, or have simply exposed the possible ramifications of militarization by saying the matter demands more study. Two studies pointed to the military as being responsible for social change among tribal groups. Lee and Hurlich, two anthropologists with extensive fieldwork among the Bushmen (Namibian San), had the following to say about the social change of the San since their recruitment by the South African Defense Forces:

> The main means for achieving this goal of a major overhaul of the social, economic, and cultural life is to combine civil and military action so that the army becomes the main agent for social change (Lee and Hurlich, 1982:336).

Dr. Zasoff, a researcher with the Rand Corporation, completed a study on the Lao Peoples Liberation Army, the communist army then fighting in Laos during the Second Indochina War. Zasloff made these observations concerning social change and the military:

> . . . we once again emphasize the importance of the LPLA as a socializing agency with the Lao Patriotic Front system . . . Having grown up in a restricted milieu that emphasizes separateness, their world revolves around their own ethnic group or tribe,

> those who speak their language or come from their region . . . Taken from a kinship group or tribe whose habits and values are determined by tradition, they are instructed to confrom to the rules of a relatively modern military organization. Some learn to read and write, some acquire a new language, and all develop certain military skills. We do not have the means to measure the total impact of this training, or indeed of the entire military experience, yet we are convinced that the PL (Pathet Lao) soldier undergoes a significant change that will, in turn, affect the larger society in which he lives" (Zasloff, 1973:89).

Whereas Zasloff felt there was at present no means to study the effects of militarization, the two anthropologists Lee and Hurlich felt that anthropology may hold the means whereby an individual may study the problem. Lee and Hurlich stated that the intent of their article was:

> " . . . to document this militarization and to show that just as history does not stand still, neither does the subject matter of anthropology come to an end when the last hunter-gatherer lays down his bow. The challenge to anthropologists is two-fold, to understand the dangerous realities facing native peoples, and also to do something about them ((Lee and Hurlich, 1982:327).

This author agrees with three points of the statement above: 1) that various examples of tribal militarization should be documented, 2) that anthropologists should continue to investigate tribal groups that have become militarized, 3) that anthropologists should understand the "dangerous realities" that face militarized tribal societies. As pertains to the last statement that anthropologists should "do something," this author is unable to suggest what should be done other than that further research into this area may enhance our interpretation of historical events.

The following thesis is designed as per the suggestions of Lee and Hurlich. Chapters I and II document the communist and Western theories of insurgency and counterinsurgency warfare, the two most prevalent forms of unconventional warfare that dictate the militarization of tribal groups. Chapters III and IV are case studies from Vietnam that investigate the historical military role(s) of tribal soldiers, primarily in unconventional modes of warfare. Chapter V analyzes not only the effects on tribal leadership and the population in general, but also the future ramifications and "dangerous realities" of the policy or arming tribal populations.

Principle Sources cited in This Thesis

During an initial review of material for this thesis, a number of books were rejected by this author for being either too vague or general for an anthropological re-analysis. Examples of such books are *Ghurka*, *Search Out the Land*, and *Ethnic Soldier*. There are also autobiographical accounts of nation-state soldiers who led tribal soldiers including T. E. Lawrence's above-mentioned book, *The Seven Pillars of Wisdom*, and Frederick Burnham's *Chief Scout*. Although these are all excellent books they are neither related to Southeast Asia, nor considered examples of modern unconventional warfare.

One of the few articles devoted to studying the role of hill tribal peoples in Southeast Asia during the First Indochina War (1946-54) is "Mountain Minorities and the Vietminh" by John T. McAlister Jr. (1967). Much valuable information was obtained on the Tho tribal people of North Vietnam and the politics of the various tribal peoples of the Tai Highlands of northwest North Vietnam. This author decided to adopt McAlister's terminology of "highlanders" and "lowlanders" to this thesis. "Highlander" is a generic term which refers to such upland hill tribal groups as the White Tai, Black Tai, Meo (Hmong), Tho, etc. The term "lowlander" denotes the ethnic Vietnamese who live in the eastern plains and coastal regions of North and South Vietnam.

One of the other few books to be written on highlanders in military service during the Second Indochina War (the American involvement from 1961-73) is *U.S. Army Special Forces* (1972) by Francis Kelly. Whereas McAlister's primary emphasis is on the highlanders of North Vietnam, Kelly's book is a history of the highlanders, or Montagnards, of South Vietnam in relation to the U.S. Army Special Forces. In order to alleviate any confusion between highlanders of North and South Vietnam, the term Montagnards will be used in reference to the highlanders of

South Vietnam. Valuable statistical information such as troop strengths of village militia forces was obtained through this book. This is possible one of the best books available in English which gives a general overview of military and para-military forces designed specifically to utilize indigenous tribesmen.

Dr. Gerald Hickey has written two masterful and definitive books specifically on the *Montagnards: Sons of the Mountains* (1982a) and *Free in the Forest* (1982b). It should be noted that Hickey received his PhD in anthropology after extensive field research in South Vietnam. During the Second Indochina War, Hickey was involved in various anthropological projects concerning the Montagnards (i.e., the Michigan State University Group) collecting linguistic and ethnographic data. His books are of special interest for this thesis in that they contain extensive biographical data on Montagnard leaders.

Hickey's books are possibly the only to contain an analysis of the various Montagnard ethnonationalist movements in detail. Biographical data and historical background information was provided on the ethnonationalist movements Front for the Liberation of Montagnards (FLM), the Bajaraka (acronym for the Bahnar, Jarai, Rhade and Koho tribes), and the United Struggle Front for the Oppressed Races (FULRO). Due to Hickey's extensive research, an anthropological re-analysis of his data by this author was possible. For this reason Hickey is heavily cited in Chapter IV of this thesis.

Dr. Bernard Fall (PhD in Political Science) was possibly the most astute and prolific writer on Southeast Asia, particularly Vietnam, until his death in Vietnam in 1967. Historical information concerning colonial and indigenous tribal units involved in highland battles was obtained from *Street Without Joy* and *Hell in a Very Small Place*. Fall and Hickey both lived in Vietnam for years and had written their dissertations after completing fieldwork in Vietnam. Having personally met many indigenous leaders, both authors were able to provide information not available to other researchers.

This author's attention was first drawn to the problem of the "parallel hierarchies" and "front organizations" of the Vietnminh and Viet Cong by Fall. The parallel hierarchies is a political system which parallels each legitimate government office with a Communist equivalent, thereby isolating the population from the control of the legitimate government. The front organizations contain the committees such as the "Peasant Association," "Youth Association," etc.

The political system of the parallel hierarchies is very little mentioned in written materials by other authors. In conjunction with the Vietminh style of warfare, the parallel hierarchies were first brought to public attention the French military magazine *Revue Militaire d'Information* (Nos. 280 and 281 March/April 1957). Fall quoted both these sources as did Arthur J. Dommen in his scholarly and oft-quoted book *Conflict in Laos* (1964, revised in 1971). Special attention to the parallel hierarchies and their front organizations has been included by this author in Chapter I and V of this thesis.

Another author who based his books on actual fieldwork is Alfred McCoy in *The Politics of Heroin in Southeast Asia* and as a contributing editor of *Laos: War and Revolution*. McCoy's books are comparable to such anti-Western authors as the Australian Communist Wilfred Burchett and the American Noam Chomsky, but he is still rather unique. Valuable information on the tribal guerrilla units know as the Mixed Airborne Commando Groups (GCMA's) was obtained through McCoy. His research included highly revealing and frank interviews with such Western and indigenous leaders as Sainteny (French administrator in the Tai Highlands) and Roger Trinquier (co-founder, director, and theoretician of the GCMA). McCoy is not without fault, as evidenced by a passage on page 91. He states that the tribesmen are faced with the decision in May whether to plant rice or opium. All other crops except opium in the highlands are planted in May and harvested in August. There is no "whether or not" decision at all, as opium is planted in November and harvested in March.

Further information on leading personalities was obtained through William J. Duiker's *The Communist Road to Power in Vietnam* (1981). Duiker, now retired, served in the U.S. Embassy in Vietnam during the Second Indochina War. His book is a very prescient political analysis of the Indochinese Communist Party (ICP) and its' founder Ho Chi Minh. As with Hickey's books on Montagnard leadership, Duiker's book includes biographical data on high-level highland Communist leaders such as Chu Van Tan and Le Quang Ba after the Fall of Saigon in 1975.

Douglas Blaufarb, formerly of the Central Intelligence Agency, is the author of *The Counterinsurgency Era: US Doctrine and Performance* (1977). With the exception of information on the Laotian Hmong guerrillas and General Vang Pao, this book is of general information and offers no new insights (see the preface of his book). Blaufarb's book should probably be read in conjunction with Geoffrey Fairbairn's *Revolutionary Guerrilla Warfare: The Countryside Version* (1974). Fairbairn, a former WWII commando, history lecturer, and analyst-at-large gives a very detailed progression of Communist unconventional (insurgency) warfare starting with Lenin and Mao before proceeding to Vietnam. It should be noted that Fairbairn is one of the few authors to write about the political role of parallel hierarchies and front organizations in conjunction with the land reform policies of Communist regimes.

Lee and Hurlich, at first brought to the attention of the reader in the introduction, are two of the very few in the field of anthropology to concern themselves with the topic of unconventional modern warfare. This author has found it surprising, as evidenced by this review of authors and material, that the vast majority of individuals researching unconventional warfare are political scientists and historians. Lee and Hurlich's article "From Foragers to Fighters: the Militarization of the Namibian San" concerns specifically the role of the San Bushmen in the continuing war in Namibia (Southwest Africa). Both authors had done fieldwork on the San and were concerned about the recruitment of the San as counterinsurgency soldiers to fight the insurgent Southwest African People's Organization (SWAPO). SWAPO soldiers are primarily Ovambu tribesmen, and Lee and Hurlich were concerned about possible tribal animosities being manipulated for the war effort.

The article by Lee and Hurlich should be viewed as a breakthrough for anthropology and warfare studies. They were bold enough to draw attention to the almost epidemic problem of Third World militarization and their approach for study was adopted by this author. Although their approach was considered sound, their observations seemed rather naive to this author. No mention was made of parallel hierarchies or front organizations, a subject deemed of considerable importance by this author in relation to any anthropological study of unconventional warfare. Lee and Hurlich did mention the starting of "weaver's associations" by

SWAPO, which they viewed as a "positive" step toward modernization. Due to the facts to be presented in Chapter V, this author does not view "weaver's associations" as positive in any way.

A variety of ethnographic materials were used for maps, tables, and legendary history. DA Pamphlet 550-110, *Minority Groups in North Vietnam* (1972) is a highly detailed and well-organized collection of ethnographic data on such tribal groups as the Tai and Meo (Hmong). DA Pamphlet 550-105, *Minority Groups in the Republic of Vietnam* (1966) is not as highly detailed as DA Pamphlet 550-110, but still contains excellent information on social structure and geographical tribal demographics. Legendary history is the only particularly notable absence of data in DA Pamphlet 550-105.

Lastly, there are a number of first-hand accounts by men who actually led indigenous units. Rene Riesen was a corporal in the French Army and the leader of a GCMA unit in South Vietnam during the First Indochina War. Riesen, who had learned a number of tribal dialects, describes his experiences among the Montagnards in lay anthropological terms. The Barking Deer (1974) by John Rubin is a novelized account of a U.S. Special Forces team on the internal politics of a Rhade village during the beginning of the Second Indochina War. Although these books are written as novels, the ethnographic data contained in them is correct when compared to the DA Pamphlets cited above.

CHAPTER I

Communist Insurgency Warfare

IT WAS DECIDED best to start this probe of unconventional warfare and Vietnam with a discussion of the lowlander Vietnamese rather than first with the highlanders themselves. More specifically, the discussion should begin, or use as a focal point, the leaders of the Vietnamese insurgency movement: Ho Chi Minh, Indochinese Communist Party founder and leader; General Giap, military leader and theorist; and Troung Chin, Party theorist. It is these individuals who would develop the political and military theories that would dictate the arming of tribal peoples in Vietnam as essential to their victory. What follows is a discussion of how Ho Chi Minh was able to effectively transform the tactics of the Chinese Communist revolution into a Vietnamese Communist revolution.

Confucianism and Communism: A Blueprint for Victory

Interestingly enough there is a certain pattern to the lowlanders who became involved in Vietnamese nationalist movements, particularly Ho's party the Lao Dong. One author quotes a Communist historian who claimed that 90% of their own rank and file came from the urban petty bourgeoisie. Most importantly for our analysis, many of the prominent leaders, such as Ho himself, came from a strong Confucian background, a qualification for political leadership in Vietnam **before** the arrival of the French colonialists (Duiker, 1981:25). Vietnamese leadership can also be analyzed geographically in that the majority of the Communist nationalist leaders came from North and Central Vietnam. This is interesting in that Vietnamese culture had not changed as much in these areas as in South Vietnam. It was in the South that the French had transformed the traditional economy into a more capitalist approach. Industrialization was begun here and Saigon was now a city of commerce with a budding new element of society, the merchant, or middle-class. A rival non-Communist nationalist movement, the VNQDD, was particularly strong in Saigon.

This geographic factor offered the lowlanders two distinct viewpoints of Vietnam. In the North, the Confucian educated Mandarins, who still exercised control, were viewed as venal and corrupt and the object of criticism as degenerated opiated bureaucrats in Vietnamese literature (McCoy, 1972:76). The Southern lowlanders, on the other hand, were also viewed with disdain by their own countrymen as degenerates due to corruption and material greed.

Mr. Duiker makes a comparison with the beginnings of Marxism in China and Vietnam. Many of the early leaders of the Chinese Communist Party (CCP) were from the rural scholar-gentry from underdeveloped areas like Central Yangtze and Szechwan. The rival Koumintang, on the other hand, came from commercial families from coastal areas that had been "Westernized." The comparison then is that the CCP:LD::KMT: VNQDD.

At first glance this may seem like an unnecessary comparison worth only a passing philosophical discussion unrelated to insurgency (revolutionary) warfare. This would be the case except for a number of important interrelated facts: many Viet Minh got their start in with the CCP and the KMT, then under Sun Yat-sen's control, and Ho's involvement in a Kremlin mission to the KMT. It was during the "Borodin Mission" that Ho would be able to gain valuable practical experience in how an Asian revolutionary movement is organized politically and militarily (Leng and Palmer, 1960:68).

Dr. Sun, one of the first true modern Chinese intellectuals and often referred to as the "Father of Modern China," had been impressed by the success of the Communist victory in Russia. He noted the parallels between China and Russia, particularly that both countries were large and considered "backward" by Western industrial standards. Dr. Sun sent the first foreign letter of congratulations to Lenin and asked for assistance for his own "revolution." The result was the Borodin Mission in which Michael Borodin was sent as an "advisor," along with Ho Chi Minh as an interpreter, to the KMT (Karnow, 1983:123).

Borodin found the KMT in disarray and set about reorganizing the various KMT factions into a tight pyramid structure. Borodin explained to Dr. Sun his organization needed more appeal to the masses and hence needed an effective propaganda effort. For military training General Chiang Kai-shek had been sent to Moscow and would later establish a military academy for the KMT at Wampoa (Asprey, 1975:340). Most importantly, Borodin integrated the two separate arms of the KMT, the political and military, into a soviet model.

An element to be further analyzed is the role of Confucianism to Marxism and "elite rule." At first glance Confucianism and Marxism may seem incompatible due to the following:

> "Confucianism has a concept of society as static. It deprecates wealth as an obstacle to high ethical standards. It is fundamentally hierarchical and accepts differences in status and capacity . . .

> "Marxism is dynamic and progress-oriented. It glories the process and views man as a natural creator. It is egalitarian and attempts to abolish differences" (Duiker, 1982:25).

In the case of China, Dr. Sun had fit Marxism into Confucian philosophy. Dr. Sun considered Marxist theory incompatible to the situation in China due to his belief that there existed no basis for "class war and the dictatorship of the proletariat" (Leng and Palmer, 1960:85). Dr. Sun did see a connection between "communism" and the Great Commonwealth of Confucius:

> "We cannot say, then, that the theory of communism is different from our Min-sheng Principle. Our Three Principles of the

> People mean government 'of the people, by the people, and for the people'—that is, a state belonging to all the people, a government controlled by all the people, and rights and benefits for the enjoyment of all the people. If this is true, the people will not only have communistic share in state production, but they will have a share in everything. When the people share everything in the state, then we truly reach the goal of the Min-sheng Principle, which is Confucius' hope of a 'great commonwealth'."

Dr. Sun's marriage of Marxism and Confucianism was not lost on Ho.[1] In the Vietnamese situation Confucianism was not enough, as had been proven by the Vietnamese Mandarin defeat at the hands of the French. The Vietnamese had seen the Mandarins degenerate from an exalted position into corrupt toadies for the French colonial administration. In the case of Ho, Marxism could be an element that could **modernize** certain Confucian principles by adding to the deficiencies of Confucianism. In this way Ho's brand of Marxism could be viewed at once as a break with the humiliating past, that is the traditional Vietnamese political structure's inability to deal with French colonialism, but at the same time the new solution could be viewed as indigenous and quasi-traditional. One author duly noted these similarities between Marxism and Confucianism with which the Vietnamese could identify:

> "The belief in one truth, embodied in quasi-sacred texts: the concept of an anointed elite, trained in an all-embracing doctrine and responsible for leading the broad masses and indoctrinating them in proper thought and behavior; the stress on personal ethics and selfless service to society; the subordination of the individual to the community; the belief that material is not the ultimate end product of human endeavor but should be firmly subordinated to more spiritual goals; and the convictions that human nature is malleable and can be improved through corrective action" (Duiker, 1982:26).

Ho also presented the Vietnamese with a leadership "role" with which they could identify. Rather than pose as a father figure, which in the family-oriented society means sternness, he chose instead the "uncle figure." The Vietnamese societal perception of "Uncle Ho" was that of a permissive individual who gives sage advice rather than an authoritative individual who demands submission.

This image would go beyond such a superficial, but effective role as "uncle" to embrace another positive element. Not content with being just a "man of the people" and "uncle figure," Ho also projected the image of the Confucian "superior man." In Confucianism it is know as *chun tzu*, but in Vietnam it is referred to as *quan tu*. A man who possessed *quan tu* had the best qualities of Confucianism: rectitude, probity, sincerity, modesty, courage, and self-sacrifice (Duiker, 1981:27).[2]

As a comparative example, South Vietnam's President Ngo Dinh Diem offered countrymen with an almost totally alien image. Diem was also Confucian-educated, but had converted to Catholicism. He tried to inspire his country with a progressive Catholic philosophy know as "personalism" to combat not only Ho's Communism, but Western liberalism as well. Personalism was conceived in 1930 primarily by the French monk Emmanuel Mounier. Considered abstract by even Western standards, it was totally incomprehensible to the Vietnamese. Ironically, Diem's brand of personalism was denounced as fraudulent by Mounier's successors (Karnow, 1983:265).

Before proceeding to the next section it is necessary to point to one last political comparison between Confucianism and Asian Communist leadership—the concept of elite rule. Ho was able to observe in China that the Koumintang and the Chinese Communist Party both shared the same opinion that the masses needed to be ruled by an enlightened group of experts. There was one major difference in the KMT envisioned ruling the masses, whereas the CCP " . . . had a more subtle program to take the masses into its fold and to ***rule through them***" (Leng and Palmer, 1960:149, italics added).

In the case of Vietnam, Ho was to expand on the fundamental principle of ruling through the masses. For this purpose he developed, or at least adapted, an ingenious system known as the *parallel hierarchies*. What should concern the anthropologist about this system is the possibility that this was the means by which a tribal political system could be slowly transformed into a Communist political system. Whereas Diem was outwardly stressing the socio-political assimilation of tribal groups into the mainstream of Vietnamese society, Ho was outwardly pledging autonomy while implementing a slow program of tribal socio-political assimilation. As will be shown, the pledge of autonomy was merely a device by which to at first attract tribal support for the Communist cause. It was the *parallel hierarchies* that would eventually ensnare the tribal society into a web it would be unable to extract itself from.

Parallel Hierarchies: Ruling Through the Masses

The form and use of *parallel hierarchies* is nothing new to the political arena in history.[3] Both the Bolshevik Revolution against the czars and the French Resistance against the Vichy French had utilized this political system.

The purpose was to place an organization's own members into an already existing political framework in order to bring the population under their control in ever increasing numbers. The Viet Minh would once again mold, or "Vietnamize" this idea to fit their own situation. It should be remembered that the Viet Minh did not control the cities and were forced to retreat into the highlands. The *parallel hierarchies* then would be used on various tribal political units to bring this section of the population under their control. The French Army completed a study on the parallel organizations and their effectiveness in 1957, from which the following was taken:

> "The Lien Viet (organization of Viet Minh subsidiary groups) included youth groups, groups for mothers, farmers, workers, 'resistance' Catholics, war veterans, etc. It could just as well have included associations of flute players or bicycle racers; the important point was that no one escaped regimentation and that the (normal) territorial hierarchy was thus complimented by another which watched the former and was in turn watched by it—both of them being watched in turn from the outside and inside by the security services and the Party. The individual caught in the fine mesh of such a net has no chance what ever of preserving his independence" (Fall, 1963:134)

It is almost unbelievable that even though the French clearly understood the nature of *parallel hierarchies*, the Americans did not even translate the French studies concerning the *parallel hierarchies.* A former Rand researcher, Zasloff, describes this system in his books on the Pathet Lao of Laos. His book appeared fully fifteen years after the first French studies were publicly printed in 1957. Zasloff does not refer to the political system he describes as a *parallel hierarchy*, but rather as the "dual front political system." Unfortunately, Zasloff does not get into any detail of how a tribal culture, specifically its own political structure will eventually become a worthless extension of the *parallel hierarchy.* As quoted in the French article:

> "In the space of a few years, the Communist Party . . . has created a territorial organization that is both highly centralized and very flexible and which creates unity of command at every level from the village up . . . Furthermore, they control the (administrative) hierarchy via the ever-present members of the Party. Thus, they are able to conduct the war as they please, without fear of the opposition or of 'deviation'." (Fall, 1963:136)

Table 1 is a diagram of the Vietnamese parallel hierarchy, known by its Vietnamese name as *Uy Ban Hanh Chinh*, or UBHC. Originally all of Vietnam was divided up into 14 zones by the Viet Minh leadership following World War II. In March 1948 these were merged into 6 *lien-khu* in order to streamline the structure. Two of these were to become "autonomous," the Tay-Bac and the Viet-Bac Zones (see below). The political structure actually follows the Soviet model, in which the committees come from locally elected People's Councils, who in turn approve the committee members from their own ranks (Fall, 1963:136).

A further discussion of the effects of the *parallel hierarchies* in two tribal areas of North Vietnam, the two "autonomous zones," is being reserved for Chapter V. What is to follow is the communist theories of insurgency warfare (people's war, revolutionary warfare, etc.). What should be noted is that the use of peasants, rather than factory workers, is an obvious deviation from Lenin's principles of revolutionary warfare.

The use of "peasants" as the "masses" in warfare was a decision that would profoundly effect all major tribal groups in Indochina.

Communist Revolutionary Warfare: Warfare Through the Masses

The theoretical framework on which Ho and his group based their plans for warfare are an extension of Maoist revolutionary thinking. As with other theories adopted by Ho, i.e., *parallel hierarchies*, village soviets, etc., this would also be tailored to fit the situation in Vietnam. The uninitiated reader might be surprised to find out that many tactics were modified after a defeat and/or reappraisal of the situation. In order to fully appreciate the military tactics employed by the Vietnam Minh, it is necessary to return to a review of China an Mao.

Due to the fact that the Chinese Communist Party was Marxist-Leninist oriented, the primary emphasis of Party organization was directed at the working proletariat. The first peasant organization was actually begun by P'eng P'ai in 1921, and Mao was to eventually become involved in similar activities in Hunan Province (Asprey, 1975:345). In 1927 Mao had noted that the peasants offered an even greater potential than factory workers for armed revolt. He wrote a report to that effect to the Central Committee, stating " . . . several hundred million will rise like a mighty storm . . . so swift and violent that no power, however great, will be able to hold it back" (Asprey, 1975:346).

Just a scant three year later the CCP would launch an abortive attack against cities in the south of China. The retreat of Mao enabled him to reflect on the weaknesses of present Party policy:

> "The campaign against the cities convinced Mao and Chu that victory must come from peasant and countryside, not worker and city. As Dr. Griffith has written, this was ' . . . the single most vital decision in the history of the Chinese Communist Party.' Mao and Chu decided this in September 1930. A refutation of Marxism-Leninism, it was also a demonstration of where real Communist leadership lay in China" (Asprey, 1975:349).

Out of this experience Mao was to form his principles for "People's War." Although Mao wrote extensively on the subject of tactics, he did not develop a coherent formula for its implementation. It is the Western analysts that developed three different phases, or a least named them as such, on Mao's writings. The following is a brief synopsis of the three phases:

> Phase I: In this phase Communist Party cadre go out into the population in order to spread propaganda and recruit. This will enable them to build up an organizational base. These teams are often referred to as "armed propaganda teams." It is hoped that with these small groups a village self-defense force can be started from which main force units can be expanded upon.
>
> Phase II: This is the guerrilla warfare phase and is only launched after base areas have been established. It is in these areas that the insurgents are the new 'government,' as they are the ones now who collect taxes, start their own various health and agriculture programs, etc. New training camps are also established to train more troops.
>
> Phase III: This is the most critical phase and one in which timing is of the utmost importance. General Giap blundered a number of times in both Indochina Wars by assuming his forces were ready for the 'general counter-offensive,' only to suffer enormous losses. Two examples of this would be the Tet Offensive (1968) and the Easter Offensive (1972). (Source: Blaufarb, 1977:3-4).

It is also important to note that Phase II can last for a number of years, as evidenced not only by the First and Second Indochina Wars, but also by many present on-going conflicts such as Angola and Namibia (see Venter, 1973).[4] During the conflict Phase I may be being waged in one section of the country, while Phase II is simultaneously being waged in another part. As such, this system of warfare is flexible. Phase II may be reached only to have to return for an interim period to Phase I when faced with heavy pressure and possible destruction by government forces.

It is the Vietnamese that were to add a new twist to the role of the guerrilla in war. Mao had written on protracted warfare and wearing the enemy down by attrition. This strategy is not new to the art of warfare and was even the underlying principle behind the Battle of Verdun when the German General Staff attempted to "bleed the French Army white" during World War I.

Ho's strategy was meant to go even further with calculated effects on the morale of the home front. To determine military strategy based on public sentiment was not Maoist in nature. Public sentiment against a war would be aroused by the continuing "body count" to the point that the war effort would seem too costly (Blaufarb, 1977:12). Having been to France before, Ho knew that the French public would not tolerate heavy losses in a "colonial" war. As he explained it to Jean Sainteny in 1946:

> " . . . If we have to fight, we will fight. You will kill ten of our men and we will kill one of yours, and in the end it will be you who will tire of it" (Asprey, 1975:xix).

In order to achieve this goal, that is the monumental human effort needed to win this style of warfare, Ho needed to instill the concept of a national identity and purpose into a diverse population. A national identity is not enough and what is most important is the fanatical determination by the population to carry the effort through to the end. What follows in the last section is the propaganda campaign and the men who carried it out.

Propaganda and the Secret Police: Controlling the Masses

In order for the Communist cadre to recruit villagers into their ranks of the armed, or in any other capacity, the villagers would first have to be enlightened, coerced, or even terrorized into following cadre instruction. Villagers could first be introduced to the Communist cause by a visit from a traveling circus or theater troupe. Skits would be performed for the villagers that would not only be entertaining, but also contain a moral. Cadre officials would be on hand to explain the significance of the plays and either ask, harangue, or cajole the villagers for recruits.

The cadre officials would at first try to be pleasant with the villagers if at all possible. It was hoped that the diligence and enthusiasm of the cadre would impress the villagers enough to create a following (Asprey, 1975:1000). Cadre officials would have to memorize the "12 Rules for Dealing With the People" (Zasloff, 1973).

"4 Respects

Respect and treat people of all ethnic groups as a father, mother, brother and sister of the same family.
Respect the traditions of people of all ethnic groups.
Respect the beliefs of the people, the monks and clergy.
Respect and support the local administration.

"4 Dont's

Don't beat, threaten, or swear at the people.
Don't take any property from the people without permission.
Don't destroy temples, churches or public property.
Don't take liberties with girls.

"4 Helps

Help people earn their living, improve their production and repair their houses.
Help educate the people about politics and culture.
Help the people improve their sanitation and give them medical aid when necessary.
Help the people fight against the enemy for the defense of the country."

It was after this initial contact phase that other cadre officials would arrive that constituted the *parallel hierarchies* as explained above. In order to insure that not only the villagers, but also the cadre would conform to official directives, an all pervasive secret police system would be integrated into the village. There were three different police organizations to deal with the enforcement problem: the *Cong-an*, *Trinh Sat*, and the *Dich Van*.

1. The *Cong-an* is the civilian secret police and operates like any other standard secret police organization, i.e., through a net of informants the police could maintain watch on villagers making disparaging remarks.
2. The *Trinh Sat* is devoted to military intelligence. One company is attached to every combat division, with smaller units being attached to the combat infantry companies. They could also be utilized as an armed reconnaissance team in which they could precede every operation by infiltrating into the areas.
3. The *Dich Van* was the most effective means of control for the NVA, and one which the West was unable to cope with. *Dich Van* mean "Moral Intervention," but their needs of control was by terror and assassination. They would follow the concept of terror as espoused by Lenin: 'The purpose of terror is to terrorize' (Sterling, 1981:8). The *Dich Van* would assassinate village chief in gruesome ways, and implement other death sentences for villagers that had committed other 'crimes against the people'" (Source: Fall, 1963:137)

The effects of such a scientific theoretical approach to Communist insurgency is being reserved for the conclusion of this thesis. What follows is the Western response to such a well-organized insurgency. It should be kept in mind at all times by the reader that the Western approach was primarily a counter-measure to the various facets of the insurgency as described above.

In conclusion, the most tangible offering the Communists could make their countrymen was the possibility of independence. The slogan "*Doc Lap*!" (Independence was a powerful incentive for the Vietnamese to face the challenge. Translated into the appropriate tribal dialects, *Doc Lap* could produce the same desired effect. But what could the West as a foreign or even colonial power offer the population in return? These and other questions will be examined in the following chapter.

CHAPTER II

Counterinsurgency Warfare

AN ACCOUNT OF the historical rise, or development of counterinsurgency theories is beyond the scope of this thesis, and would also uselessly clutter the final analysis. A much easier approach would be to begin with the counterinsurgency techniques as developed and applied in Malaya and the Philippines. The Philippine counterinsurgency effort is one of the examples of counterinsurgency which greatly influenced the thinking of President John Kennedy in our own involvement in Vietnam. It should also be noted that Sir Robert Thompson, a high-ranking British official during the Malayan Emergency was a special advisor to President Lyndon Johnson during the Second Indochina War.

Malaya and the Philippines: The Basis for American Theory

The Communist insurgency in Malaya began in 1948 at the instigation of the Communist Party of Malaya (MCP). The MCP had only begun to organize circa 1930, but the Japanese invasion of Manchuria was a catalyst that brought the MCP in contact with the KMT of China during the period of 1936-37 (Asprey, 1975:577). The MCP received arms from the Western powers during World War II and formed a guerrilla army called the Anti-Japanese Union and Forces (MPAJUP). This force remained relatively intact after the war, possibly due to the fact that time was spent organizing rather than fighting the Japanese (Asprey, 1975:583).

The leader of the Communist insurgency was Ch'en P'ing, a former anti-Japanese guerrilla leader who had been awarded the Order of the British Empire. In 1947 he converted his former guerrilla army (the MPAJUF) to the task of overthrowing British rule, calling his new army the "Malayan People's Anti-British Army." In the rural areas he instituted parallel hierarchies at the village to the province levels. In the cities a terrorist group called the "Blood and Steel Corps" was formed to foment unrest (Asprey, 1975:782-3).

Ch'en P'ing was quite successful at the beginning of his campaign. The British at first attempted to suppress the insurgency by utilizing only conventional troops, a tactic which failed to achieve any notable success. After 1950 the British began utilizing unconventional units, a decision which brought a curious ethnic composition of British soldiers to Malaya. The British gained the offensive with the aid of such units as the Nepalese Gurkhas and the Sarawak Rangers, a unit formed by the British of cannibalistic Iban tribesmen from Borneo. The "Emergency" was quelled in 1955, but Ch'en P'ing was never captured and is still active on a much reduced scale.

At the end of World War II the Philippine Government became involved with a guerrilla war against communist insurgents. The leader of the insurgency was Luis Taruc, the guerrilla leader of the communist "Hukbo ng Bayan so Hapon," the People's Army to Fight the Japanese. For the sake of brevity the insurgents were referred to as the "Huks" (Asprey, 1975:520). Taruc was originally given a seat in the Philippine Congress after World War II, but the communists were shortly disqualified from the government (Asprey, 1975:747).

When Taruc retreated into the hills and called for a revolt, the Philippine Government over-reacted militarily. The government re-instituted the Japanese practice of the "zona," which was the "re-settlement" of villagers. This was usually achieved at gun-point and the soldiers often abused the villagers and looted their property. The villagers naturally saw the soldiers as representatives of the legitimate government, and the soldiers would have to gain the respect of the villagers if the government wanted to control the rural areas.

The man who would rise to save the government was Magsaysay, who had been a guerrilla leader during the war and had led a guerrilla army of 10,000 men (Blaufarb, 1977:29). He was named to the post of Secretary of National Defense and first set out to straighten out the army. He turned the army into free-lance goodwill ambassadors to the villagers. Chewing gum was given to the troops to give to the children and the troops also carried more rations than they needed in order to give their surplus to needy villagers.

These and other practices were eventually to become institutionalized into the civil Affairs Office under the army. One other simple solution to the problem of official corruption, not to mention a good means of intelligence, was that the villagers could send a telegram directly to Magsaysay's office for five cents to report any violations. Most importantly, the information received would be acted upon. The army engineers would also prove valuable in counterinsurgency. Schools would be built (prefab models) and staffed with teachers. If no civilians were available for an area, the army would provide for teachers out of their own ranks.

John Kennedy, then a senator, was familiar with Magsaysay's approach to stemming communist insurgency warfare, and had even visited Vietnam in 1951 for an assessment of the war effort by the French. Senator Kennedy, aside form such authors as Blaufarb, were not impressed with the French counterinsurgency efforts, many of which were rather novel, to stop the insurgents. Peter Paret, a very prescient observer and author of *French Counter-Insurgency Warfare*, noted the primary deficiency in French theory and practice was that no new political structure was being built up (see below).

Blaufarb, among others, gives very short shrift of the French equivalent of counterinsurgency warfare, and refers to the French school of thought as *la guerre revolutionnaire*. According to him:

> "It was a somewhat frantic and intense Gallic effort to generalize from French experience to justify a world wide approach which duplicated the mechanics of the Mao/Giap system while ignoring its political substance" (Blaufarb, 1977:49, italics added).

But there are two very important facts that Blaufarb forgot to mention: 1) modern warfare was designed to operate at the rear of the enemy and interdict supply lines, etc. As such it was primarily offensive in character and not a defensive plant (Trinquier, see below), and 2) the CIA and the US Army were offered all GCMA units at the end of the First Indochina War, which they initially refused. A few years later American representatives approached Trinquier and even offered control of counterinsurgency programs in Vietnam during the Second Indochina War, which he in turn refused (McCoy, 1972:107).

Trinquier does not even attempt to hide the offensive characteristics of his, what he himself termed *la guerre modern*:

> "Attached on our own territory, we must first defend ourselves. Then we may carry the war to the enemy and grant them no respite until they capitulate. We will attack them on their terrain with weapons of modern warfare that will permit us to strike directly, in their territory, without exposing ourselves to the international complications the employment of traditional arms would surely evoke" (Trinquier, 1964:105).

What follows in the next two sections is a discussion of French counterinsurgency theory as espoused and carried out by its founder, Colonel Roger Trinquier. Trinquier envisioned matching Maoist revolutionary theory with his own. A "tribal uprising" would be matched by another "tribal uprising," a practice that would set in motion a policy of war by proxy using armed tribesmen on both sides. What should be kept in mind is that many of the tribesmen who served with the French would later serve with the US Special Forces, thereby enabling us to view culture change over an extended period.

Trinquier and Modern Warfare: Tribal Manipulation

French counterinsurgency techniques are the product of the military theories of Colonel Roger Trinquier. Before a discussion of Trinquier himself, it should be noted that a French political condition may have been ultimately responsible for the formulating of Trinquier's theories. Dr. Fall points to the lack of political leadership on the part of the Fourth Republic as a possible cause to a certain rise of French officer behavior:

> "Local commanders, for example, had to make the decision whether or not to arm local levies and if so, of what political or religious persuasion. In Indochina, such officer—often of captain's rank or lower-raised Catholic, Buddhist, Cao-Dai, Hoa-Hao, or mountain tribal militia forces whenever they did not use outright river pirates or deserters from the Communists" (Fall in Trinquier, 1964:viii).

One such officer who was forced to make such decisions even prior to formulating his own theories was Roger Trinquier. Showing promise in school, he was to eventually graduate with a teaching degree and planned on returning to the countryside and be a teacher. He first had to serve two years of compulsory military service. Since the backbone of the French Officer Corps is considered the teacher, Trinquier became an officer in 1931 in the French Marine Infantry.

It seems that Trinquier found his "niche" in society in the army.[5] His first duty station was at Chi-Ma on the border with Vietnam and China. His job was to fight river pirates and opium smugglers, which gives a logical explanation to his future Operation X (see below). He also showed linguistic talents during his first tour by learning some tribal dialects. Later he learned Chinese while serving as an Embassy guard in Peking in the late 1930's, which would have enabled him to have read an original Mao handbook. At the start of the First Indochina War in 1946, he returned to Indochina and served as a commando. During his second tour he served as second-in-command to the 1st Colonial Parachute Battalion.

It was during his third tour of duty in Indochina that Trinquier would eventually assume command of what was called the GCMA.[6] It was possible the then commander in Indochina, General de Lattre, who suggested using the Viet Minh's own tactics on an experimental basis. Trinquier was chosen to command this experiment due to his past experiences in Indochina. Intererstingly, it was some American military advisors in Saigon who had heard of Trinquier and the GCMA and had him visit an American anti-guerrilla school in Korea. Two young American officers, whose names were unfortunately not provided, returned with Trinquier to Indochina in order to receive first-hand knowledge of his operational techniques. American equipment was also being provided for his operations, which would suggest that the American military establishment did not totally dismiss his concepts.

During 1950 it was decided to proceed with the arming of tribesmen for the GCMA and a special school was set up for them, the Action School in Cap St. Jacques (Vung Tau). Although McCoy (1972:98) alludes to the fact that this was the first time that tribesmen would be employed as "mountain mercenaries," as he constantly refers to them, the reader will be acquainted later with the fact that tribesmen had been serving in the French Army for years. The famous "Montagnard Division" was already operational, not to mention the lesser units.

This information concerning the author McCoy is provided in order to prepare the reader for the following information about the GCMA. One of the very few sources of information on the GCMA is from McCoy, who also interviewed Trinquier concerning his Indochina activities, a rarity in itself. Trinquier developed a four stage plan for the recruitment and training of what he himself referred to as *maquis units*, a term derived from the French Resistance of World War II. Trinquier provided the following to McCoy by reading from a training manual he had written for the French Army:

"Preliminary Stage. A small group of carefully selected officers fly over hill tribe villages in a light aircraft to test the response of the inhabitants. If somebody shoots at the aircraft, the area is probably hostile, but if the tribesmen wave, then the area might have potential. In 1951, for example, Major Trinquier organized the first maquis in Tonkin by repeatedly flying over Meo villages northwest of Lai Chau until he drew a response. When some of the Meo waved a French tricolor, he realized the area qualified for state 1.

> Stage 1. Four or five MACG commandos were parachuted into the target area to recruit about fifty local tribesmen for counterguerrilla training at the Action School in Cap Saint Jacques, where up to three hundred guerrillas could be trained at a time. Trinquier later explained his criterion for selecting these first tribal cadres: . . . These ambitions mercenaries were given a forty-day commando course comprising airborne training, radio operation, demolition, small arms use, and

> counterintelligence. Afterward the group was broken up into four-man teams comprised of a combat commander, radio operator, and two intelligence officers. The teams were trained to operate independently of one another so that the maquis could survive were any of the teams captured. Stage 1 took two and a half months, and was budgeted at $3,000.
>
> "Stage 2. The original recruits returned to their home area with arms, radios, and money to set up the maquis. *Through their friends and relatives,* they began propagandizing the local population and gathering basic intelligence about Viet Minh activities in the area. Stage 2 was considered completed when the initial teams had managed to recruit a hundred more of their fellow tribesmen for training at Cap Saint Jacques. This stage usually took about two months and $6,000 with most of the increased expenses consisting of the relatively high salaries of these mercenary troops.
>
> "Stage 3 was by far the most complex and critical part of the entire process. The target area was transformed from an innocent scattering of mountain villages into a tightly controlled maquis. After the return of the final hundred cadres, and Viet Minh organizers were assassinated, *a tribal leader 'representative of the ethnic and geographic group predominant in the zone' was selected,* and arms were parachuted to the hill tribesmen. If the planning and organization had been properly carried out, the maquis would have up to three thousand armed tribesmen collecting intelligence, ferreting out Viet Minh camps and supply lines. Moreover, *the maquis was capable of running itself,* with the selected tribal leader communicating regularly by radio with French liaison officers in Hanoi or Saigon to assure a steady supply of arms, ammunition, and money" (McCoy, 1972:98-9, italics added).

To recapitulate the above information, Trinquier's purpose was to polarize the tribal situation. As revealed in "Stage 3,' this system was self-perpetuating, which explains the fact that the number of tribal GCMA troops can only be estimated. (The estimates vary radically from twenty to forty thousand tribal commandos.) As will be shown in the following section, Trinquier, just as the Viet Minh, also believed in mobilizing a broad base including the criminal underworld in order to support his counterinsurgency effort. Trinquier's theory and practice of "modern warfare" would have implications for not only tribal groups, but also Western nations (see below).

Operation X: Into the Milieu

In order to combat, or off-set, the Viet Minh strategy of mass uprising, Trinquier, as explained above, planned to mobilize every conceivable facet of Indochinese society. Every individual that held a grudge against the Viet Minh, and every group, cult, sect, etc., (Cao Dai, Hao Hoa, Catholic, etc.) would be armed and trained to resist. The oft heard remark that the French would "divide and rule" is not entirely true. Coalitions, however tenuous, would be attempted within the framework of the GCMA (see below).

Trinquier was faced with a much more monumental task before he could implement *modern warfare* on a grand scale—the acquisition of funds. The French Army's efforts in Vietnam were slowly being thwarted by the French National Assembly in that funds were being reduced for the war. The French Army, just as the US Army would be in the Second Indochina War, was hopelessly addicted to conventional warfare thinking. The French Army was understandable reluctant to tap into an ever dwindling monetary supply for experimentation.

The solution Trinquier and Captain Savani[7] arrived at was unorthodox and simple as their techniques in *modern warfare*. The actual acquisition of funds for their operations were known as Operation X, the funds of which would be deposited in the "black box," or *caisse noire* in Vung Tau. Indigenous commando leaders would thereby be able to personally draw money out of the black box to fund their own units.

In short, Operation X was the codeword for the complete take-over and perpetuation of the opium/heroin trade by the officers of the GCMA and SDECE (see Table 2). Although at first it may seem incredible that

French officers would be military tacticians by day and drug dealers by night, Operation X was also a very practical solution to a multifaceted problem:

1) The Viet Minh were buying opium to trade for arms, etc.
2) The French were unable to stop the growing of opium since it provided money to the tribes.
3) The French would instead buy the opium at a better price, thereby depriving the Viet Minh of a much needed source of revenue.
4) Rather than destroy the opium, the French would refine it and sell it at a profit.
5) Through the profits obtained through the opium trade, funds would be provided for tribal military operations.

With the advent of Operation X the race was on for control of opium production between the French and Viet Minh. The annual crops of opium were worth millions of dollars. In 1947, for example, 38 tons of opium grown in North Vietnam were marketed for sale. This crop was worth 400 million piasters ($16 million in 1957 dollars). Comparatively, 233,000 tons of rice exported in 1948 was worth 452 million piasters. By controlling just the fill tribes in North Vietnam, particularly the opium-producing Hmong, the Viet Minh could fiscally compete with the French Colonial Government (McAlister, 1967:821).

Opium, even in raw form, was a highly prized trade item. Near Cao Bang on the Sino-Vietnamese Border, a barter and trade network was established between the Viet Minh and the Chinese Communists. A light machine gun and 500 rounds "cost" six kilograms of opium, 2.5 kilograms "bought" a rifle and 500 rounds. Thirty-eight tons of opium through the black market were worth 12,800 rifles and 6,400,000 rounds of ammunition, or enough to arm a full division (McAlister, 1967:822).

The importance of Operation X should be put in full and proper perspective. Operation X was designed to work within an existing economic framework, albeit illicit economy, the economy of opium. Just as opium would be traded by the Hmong and other hill tribes for prestige items, primarily silver bars, the French would pay their hill tribe commandos in silver bars.

There is also a military aspect of the GCMA and Operation X, a fact that is overlooked by many authors. Just as conventional military operations would be designed to protect inductrial and civilian population centers, the GCMA and Operation X would perform a dual function in the best tradition of conventional military thinking: the protection (and thereby depriving the enemy) of a valuable hill tribe industry.

The end of the First Indochina War would bring about tragic consequences not only for the tribal groups allied with the French through the GCMA, but also for the French officer corps. The Viet Minh had made sure that the French tribal commandos were not covered under the 1954 Geneva Accords, and turned their attention to wiping out the GCMA's. The Viet Minh took relish in this task, as evidenced by their publishing a "body count" of killed and captured GCMA troops (Fall, 1961:276).

This was to take some effort. The French High Command had radioed all GCMA units to retreat to Laos or South Vietnam. The tribal soldiers, except for a few units, refused to leave their families and tribal groups unprotected against the Viet Minh. Most Frenchmen refused to leave their tribal soldiers for varied reason; some had married into the tribe, and others felt they had "given their word" to protect the tribes and as such felt they should remain to the bitter end. Even without supplies, the last GCMA unit in North Vietnam was not wiped out until two years later, with the last radio transmission being received in 1956 (Fall, 1961:278).

The Buon Enao Experiment: Winning the "Hearts and Minds"

With the refusal by Major Trinquier to head the new American effort at counterinsurgency warfare, the Americans would receive help from the British. Sir Robert Thompson, a high-ranking official during the Malayan Emergency, described earlier, would become a special advisor to the President. Also not to be left unmentioned is the fact that the Philippine Government sent a special rural development detachment, along with their own security troops, to aid the South Vietnamese Government. Needless to say, this would all add to give a certain British-Philippine flair to the American effort to gain the support, or "hearts and minds" of the populace. This Americna effort in the Central Highlands of Vietnam would bear a marked similarity to the British and Philippine counterinsurgency programs already mentioned in this chapter.

With all due respect to Sir Robert Thompson, the arming of tribesmen was somewhat of a *laissez faire* program. The Rhade village of Buon Enao, with a population of 400, was chosen as the first village in South Vietnam to be armed, and negotiations wiht the village chief was begun in the Fall of 1961. The "Buon Enao Experiment," as it was officially known, was begun in order to investigate the feasability of the tribesmen playing a more active role in their own defense.

U.S. Special Forces and Vietnamese Special Forces (LLDB), of which most were Montagnards already in the South Vietnamese Army (ARVN), began the negotiations and training. It is important to note that the Montagnards in the LLDB had already fought with the French Union Forces in the First Indochina War and had been "integrated" into the ARVN (see below). Village recruits were divided into two categories: 1) strike force, full-time soldiers who would remain in a base camp and perform district security plus act as a back-up reaction force, and 2) village defenders, part-time militiamen who would be trained and sent back to their respective villages to act as village security. Except for the training period of two to six weeks, only the strike force soldiers were paid a regular salary, and all village defenders served as volunteers.

This program was to expand rapidly under the control of the U.S. Special Forces. More and more village chiefs were to ask for the program among the Rhade as shown by the following statistics: (Kelly, 1972:29)

December 1961 to April 1962

40 village complex
Population in defended villages: 14,000
Population armed and trained: 300 strike force, 975 village defenders

April 1962 to October 1962

200 village complex
Population in defended villages: 60,000
Population armed and trained: 1,500 strike force, 10,600 village defenders

By December 1963 there were 18,000 strike force soldiers and 43,375 village defenders (later called hamlet militia) in the program. The civic action programs were also carried out by Montagnards who had been U.S. trained, but it is unknown at this time if they were always from the same tribe or area. Under the 5th Special Forces Group from 1964-70, the projects listed in Table 3 were begun and completed for, and in large part by, Montagnards.

It is quite obvious that such a large and all-encompassing social development program, albeit initiated and our by the army, would ultimately effect the tribal population. It should also be pointed out that the primary economic goal of such programs was to boost tribal food production with the introduction of modern agricultural techniques (Thompson, 1966). This stands in contrast with such peacetime rural development programs as high dams, factories and farms geared for export production.

Before moving on to the next section, which in actuality is an extension of the discussion of the Buon Enao Experiment, it may be necessary to remind the reader of the French counterinsurgency effort. The system of "screening" or "vouching" of tribal soldiers was basically the same for the French and Americans: the village chief did the vouching. The number of armed tribesmen in the Central Highlands by 1963 is also an interesting comparison to armed tribesmen in the GCMA in North Vietnam in 1953. It is also true that the French had built schools for various tribal peoples, as for instance the Ecole Sabatier for the Montagnards in Dalat.

It is also no accident that many French-educated tribal leaders would become involved in the American counterinsurgency effort. It should have been no surprise for the Americans that many of the Montagnards who showed up for hamlet militia training had already served in the French Army (see Chapter IV). But to many Americans it was a surprise some were receiving checks from the French Government for military services rendered.

Revolutionary Development and CAP's: Fighting Fire With Fire

In the mid-1960's the American advisors (in conjunction with the South Vietnamese Government) felt the best way to combat insurgency in the countryside would be to improve village life. In actuality, this may be somewhat of a break, but still within the theoretical framework, of the advice of Sir Robert Thomspon. The "British School" of counterinusrgency felt that a strong bureaucracy, or *full-time permanent public servants*, should never be overlooked in any "hearts and minds" campaign (Blaufarb, 1977:225).

The Americans deviated from this sound thinking by preferring a cadre approach. It may be safe to suppose that the cadre approach may actually be due to American "culture," that is a "quick fix" being adopted rather than a slow methodical approach to build up a stronger government in the countryside. The program was known as the Revolutionary Development (RD), also the "Real New Life Hamlet" (*Ap Doi Moi*).

The man put in charge of the training of the RD cadre was Major Nguyen Be who, not surprisingly, was a former Viet Minh officer. A training facility was established at Vyung Tau and selected cadre would undergo a thirteen week course. This training used such VC/NVA slogans as the "Three Withs: eat together, live together, and work together." Contact with the people was stressed, and as stated in the RD manual:[8]

> " . . . each cadre should have at least three families that love him and treat him as one of their relations. Cadres should stay long to help the people and never disturb them. Cadres should do, enjoy later, and *win the hearts and minds of the people*" (Blaufarb, 1977:228, italics added).

The RD cadre had at least 98 projects they had to complete, i.e., medical, security, education, counter-propaganda, etc. It is unknown at this time what exact fundtions many of the RD cadre performed in conjunction with the local government. It is known that they "assisted" the chiefs and assumed responsibility for village security. In view of the fact that a former Viet Minh officer was in charge of training, it could be assumed at this time that the RD cadre performed many of the same duties as the UBHC's. Most importantly (see Chapter V) this program was not integrated full-time into the indigenous political structure as a communist *parallel hierarchy*.

Somewhat akin to the RD's, and also begun about the same time, was the Combined Action Platoon's (CAP's). This consisted of a 14-man

Marine platoon that would be assigned to work with, aside from live in, a hamlet or village. Volunteers were screened to make sure they could live and work with Vietnamese. At its height there were 114 CAP's operating in I Corps, which was the northern area bordering on the DMZ with North Vietnam (Blaufarb, 1977:256).

An element of such programs that will be discussed later is the paramilitary forces, or militia, that were trained to take over local security. These were the Regular Forces (RF) and Popular Forces (PF). These troops could aptly be described as the "friendly" equivalent of the Viet Cong. This meant fighting guerrillas with guerrillas, or one of the few real attempts to build up an unconventional army to fight another unconventional army.

Such measures as described above were to have many positive effects on the villagers, for example the ending of "sweeps" which could lead to such incidents as My Lai (see West, 1973). Such measures also alleviated the insecurity caused by Viet Cong terrorist attacks, such as kidnaping and assassinations. The main target of Viet Cong assassinations would be the competent chief, whereas the incompetent chief was left as "proof" that an American or South Vietnamese program was doomed to failure. What follows is a section on some of the more unnecessary programs.

Phoenix and the CTT's: Losing the Hearts and Minds

The Phoenix Program was developed in order to eliminate the Viet Cong Infrastructure (VCI), a trade word referring to the Viet Cong parallel hierarchies (among other individuals) discussed in Chapter I. This was also another suggestion by the British Mission, namely Sir Robert Thompson, based on the British experiences in Malaya (Blaufarb, 1977:245).

The Phoenix Program, known also a *Phung Hoang* (all-seeing bird in Vietnamese), was initially a practical solution to an obvious problem. Unfortunately, it would degenerate into a sort of horror story for the villagers. It was initially designed to work in conjunction, or at least as an extra measure, with the RF's, PF's and RD's examined earlier (Blaufarb, 1977:248).

The VCI would be identified and "neutralized" by various means, and to "terminate with extreme prejudice" was a euphemism for assassination.

VCI and collaborators, providing three witnesses swore they were such, could be arrested and jailed for up to five years without trial. This sentence could also be extended for another five years.

This program was no doubt an evolution, or extension, of an earlier attempt to "fight fire with fire," the Counter Terror Teams (CTT). These teams were used to spread terror in the enemy controlled areas by kidnaping and assassinations. As with the above-mentioned programs, this was put under the control of the province chief, which may be a main cause leading to the abuse of power in the form of personal vendettas. If a villager or village chief became too "rich" due to an economic program, a province chief could label him a VCI and have him removed. The province chief could then assume control of the business (see Chapter V).

The reader may at first perceive the theories of communist insurgency and Western counterinsurgency as being exceptionally similar. As Trinquier proposed, and other Western military theorticians concurred, if tribal groups could be motivated for a "pro-communist" cause, then other groups could equally be motivated for an "anti-communist" cause. In this pattern, communist insurgents are seen as "arsonists," the insurgency as a "fire," and the counterinsurgents as "firemen" waiting to react when the alarm rings.

Such simplistic perceptions did not prevail among all who have studied the subject of unconventional warfare. Such authors as Asprey, Fall, Fairbairn, and others properly identified communist insurgency as "Revolutionary Warfare." One author (Fall, 1963:349) presentas a quasi-mathematical formula for the understanding of Revolutionary Warfare. In the formula, slightly modified by this author, RW::Revolutionary Warfare, PW::Psychological Warfare, and GW::Guerrilla Warfare. The formula for revolutionary ware is thus written:

RW::(PW plus GW)

With the exception of some of the above-mentioned authors, the *integrated* variables of PW and GW were either ignored or misunderstood by Western counterinsurgency specialists. It is the system of the *parallel hierarchies* which would determine the placing of the variable in parenthesis. The effects of a full-time *parallel hierarchy* was totally ignored by the West. By using the same RW formula, Western counterinsurgency could thus be written as:

COIN::GW

This formula will be enhanced as this thesis progresses, but let it suffice for now to state that the Western counterinsurgency specialists perceived only *the arming* of various tribal factions as a sufficient countermeasure. Propaganda and/or motivation of tribal soldiers is non-existent for all practical purposes. The tribesmen in Vietnam (and no doubt other wars) who either volunteered or were recruited for counterinsurgency forces were considered mercenaries by their benefactors with no political aspirations of their own. At best, the political aspirations were considered secondary to the ultimate (and elusive) goal of "victory." The communists, through the parallel hierarchies and front organizations (PW), would instill a sense of destiny in the tribesmen and refer to them as comrades and gallant freedom fighters. What follows in the next two chapters are specific examples of RW and COIN as practiced in North and South Vietnam during the First and Second Indochina Wars.

CHAPTER III

Case Studies From North Vietnam

AS EXPLAINED IN Chapter I, the Viet Minh would need the cooperation of the highland minorities for a successful campaign of a "People's War." The first tribal groups contacted and won over by the Viet Minh were situated in the north of North Vietnam. As will be shown it was a windfall for the Viet Minh to find their first "converts" among the Tho.

The major ethnic highland groups in the North come under the heading of "Tai." The major groups are the White Tai, Black Tai, the Tho, the Tai-speaking Nung, and the Nyang. As many tribal groups in Asia, these groups, particularly the Black and White Tai, did not recognize any nation-international borders. Many tribal groups maintained contact with related kin groups in Laos, Burma, and China.

There is one other problem, or phenomenon, when studying tribal groups in Asia and particularly Indochina. This is the fact that although tribal groups can be shown geographically on a map, this only shows the horizontal disposition of the tribal groups. When referring to the maps many tribal groups may seem to occupy the same geographical area, especially in northwestern North Vietnam. The problem is that there is a vertical disposition of many tribal groups. For example, the Black and White Tai inhabit the upland valleys, about halfway up the mountain live such groups as the Yao and Khmu, and at the very top of the mountain live the Hmong (Meo).

What follows is a discussion of the Tho, the largest ethnic minority in North Vietnam. It is also a moot question as to whether or not the Viet Minh would have survived politically or militarily without the assistance of the Tho.

The Tho

The Tho were one of the first groups to arrive in Vietnam and are also one of the most culturally advanced still residing in North Vietnam. Their migration southward out of China was not all at once, but had lasted for about 2,000 years. At one time they had been part of the Nan Chao kingdom in China, which as a part of the Chinese Empire, but eventually this semi-autonomous kingdom was conquered and assimilated by the Chinese in the 13th Century AD.

One important incident that should not be overlooked is that all the Tai in the Sikiang Basin in Southern China had formed their own coalition know as the *Dai Nam* two centuries before the fall of Nan Chao. The important point is that the leader of the *Dai Nam* was a Nung chief. But, as with other kingdoms through history, this one also disappeared (Schrock, 1972:452). The region the Tho would choose as their homeland in North Vietnam turned out to be of strategic importance. Unfortunately for the Tho, the area was a natural invasion route for the Chinese in their attacks on the lowland Vietnamese.

In the 15th Century there was rebellion of the Tho in North Vietnam against the Vietnamese Le Dynasty due to taxation (Schrock, 1972:452). After the revolt, the Vietnamese sent in their Mandarins to rule in order to gain better control over the region. In the 16th Century the lowland Vietnamese Mac family led an abortive putsch against the then-ruling Trinh Dynasty. The Mac family fled to the mountains, specifically the Tho homeland, and established a stronghold at Cao Bang near the Sino-Vietnamese border.

It was due to the Mandarin class, not to mention a proximity to the lowlands, that an interesting brand of Vietnamization would tack place. The mixed-blooded descendants of the lowland Vietnamese Mandarins and the Tho, called the *Tho-ti*, were to become hereditary aristocrats. The Tho-ti functioned politically as a mandarinate among the Tho and the Tho-ti conformed to the same political structure as in the lowlands (Hickey, 1958:33).

The Tho-ti held power over the Tho for a number of reasons, not the least of which is prestige. Although the rice fields were privately owned among the Tho, the Tho-ti were the only class among the Tho who could perform the proper rituals necessary for a good harvest. Most importantly, the Tho-ti performed the function of the "cultural broker" between two initially diverse cultures, the Tho and the lowland Vietnamese.

The Tho-ti as a highlander elite should be compared to other highlander elites (see below). The Tho-ti (along with the Mandarin class in general) had been disenfranchised by the French of the hereditary positions, which stands in contrast to the French treatment of the White Tai. The embitterment of the Tho-ti towards the French rule was reflected by the vast majority fo the Tho population. The Tho-ti were also the only Vietnamese-speaking highland elite, and when the Tho decided to throw in their lot with the Viet Minh, the role of the Tho-ti as invaluable cultural brokers would not be overlooked. What follows is an analysis of the military contribution of the Tho in the First Indochina War.

The Tho and the Vietnam People's Army

The Tho were to serve in the Viet Minh from simple front line infantrymen (*Bo-Doi*) all the way up to divisional and regional commanders. Of these none figure more prominently than Chu Van Tan in the Viet Minh struggle for power. Although he is described by Fall (1963:145) as a Tho chieftain, in actuality he was a Nung that operated in the Tho area as a guerrilla leader. He was already known at the time of the Bac Son Rebellion and was leading his won guerrilla force. His forces were also one of the few to escape annihilation at the hands of the French in 1941, which is a credit to his generalship. Chu Van Tan's "National Salvation Army" was primarily responsible for the "liberation" of the Viet Bac Zone (northern North Vietnam) in the early 1940's in conjunction with Giap's armed propaganda troops.

The Tho not only provided military muscle for the Viet Minh, but may also be responsible for acculturating the lowland Vietnamese to the finer art of guerrilla warfare. General Giap, before he met Ho Chi

Minh, had met the Tho leader Hoang Van Thu who informed him for the first time on their aspects of guerrilla warfare (Duiker, 1981:71). Although Duiker builds a strong case for the argument that Ho himself was responsible for the development of Vietnamese guerrilla ware per Maoist theory, some credit must obviously be given to the Tho. While Giap only had armed propaganda units operating in 1941, Chu Van Tan's army numbered over 3,000 battle-tested troops.

It was also Chu Van Tan and his Tho guerrillas who helped Ho and Giap in their abortive proclamation of independence and attack on Hanoi. The Viet Minh were forced to flee the cities by the French and the area they chose as their refuge was the haunts of the Mac family, the Tho homeland. But whereas the Mac family had chosen Cao Bang as their stronghold, Ho chose Thai Nguyen, a much more inaccessible area surrounded by swamps. Chu Van Tan, for his help in the revolt, was named Minister of Defense. A short time later he was promoted to General and Commander of the Viet Bac Zone. He was eventually to be promoted to Central Committee Member, a very prestigious position for a minority, which made him one of the most powerful leaders in the Viet Minh.

Just by sheer statistics alone the Tho were to be a major asset in the military forces. By the end of the First Indochina War there were 20,000 Tho serving in the regular forces. They would comprise two of the VPA's (Vietnam People's Army) best divisions and were led by one of their own generals, Le Quang Ba of the 316th Division. Both of these Tho divisions, the 312th and 316th, would distinguish themselves during the First Indochina War and took very heavy casualties, especially the 316th at the Battle of Dien Bien Phu. The Tho constituted fully one-fifth of the Viet Minh's regular troops in the VPA, and the number of Tho in the two divisions represented approximately five per-cent of the total estimated Tho population. The Tho soldiers would prove to be a crucial element in the Viet Minh's first full-scale battle against French regular troops on Route Coloniale 4 (R.C. 4).

The Battle of R.C. 4: End of a Colonial Era

The French had remained in strength on the northern Sino-Vietnamese border in order to interdict, or at least impede, the flow of arms and assistance from China. The advances of Mao's armies to the border in 1949 was viewed with alarm by the French Command and it was decided best to evacuate the French border garrisons.

The battle began in September 1950 with the Viet Minh units probing the French defenses. The French realized that they were facing a well-equipped regular force, a force with had undergone three months of training by Mao's army at Yen Shan in Yunnan, China. The French garrisons at Cao Bang and Lang Son (6,000 men) began retreating along R.C. 4. Progress was slow as the French commander had decided it best to bring along his heavy equipment, a decision now viewed as catastrophic.

The Tho were to play an integral part in the annihilation of the French garrisons by the Viet Minh. The 174th and 209th (Tho) Regiments were reinforcing the 308th Division, a division that was composed mainly of lowland volunteers from Hanoi, Phuc Yen, and Vinh Yen (McAlister, 1967:798). The two Tho regiments were particularly effective when the battle shifted from the roadside to the mountains when the French gambled on a desperate escape. The 209th, even after sustaining severe casualties, wiped out one pocket of French survivors near Dong Khe, while the 174th was able to overwhelm an elite French paratroop relief force at That Khe.

The French Army, not to mention the French public, was shocked after realizing the magnitude of the Battle of R.C. 4. The psychological aspects of the defeat must have been particularly devastating as this was the worst French colonial defeat since the Battle of Quebec (Fall, 1961:28). More disturbing was the reality that the Viet Minh had effectively progressed to Phase III warfare (see below). With surprising tactical skill they had employed and effectively coordinated infantry and artillery units together, some twenty battalions in all.

In relation to the Tho, the Viet Minh had also successfully bridged the traditional highlander/lowlander animosities:

> "The Battle of RC 4 showed that the Tho had been politicized and militarily well integrated into the Viet Minh. Previously the Vietnamese had only developed Vietnamized elite such the Tho-ti, but now they were assimilating them and offering them

social mobility based on military and political talents. Through this social and military mobility the Tho were able to operate over wide geographical areas . . ." (McAlister, 1967).

The final and dramatic defeat of the French forces would take place four years later at Dien Bien Phu. What follows is a discussion of the highland political and military policies that led to the historic and final confrontation in the Tai Highlands at Dien Bien Phu.

The Tai Highlands

Whereas the Viet Minh were promising autonomy to the highland areas if they would join them in their fight against the French, the French had already granted one area of North Vietnam its autonomy. This area in northwest North Vietnam was known as the Sip Song Chau Tai. Situated in an area known as the Tai Highlands, the Sip Song Chau Tai was a conglomerate of ethnic groups such as the White Tai, Black Tai, Hmong, etc.

When the French penetrated this region in the 1880's, they approached the White Tai, the *de facto* rulers of the Sip Song Chau Tai as a means to check Siamese expansionism and keep it west of the Mekong River. Not too long before French contact some areas of the Tai Highlands had become a refuge for not only the Hmong, but also for bands of Chinese such as the Black Flag pirates who had fled after the abortive Taiping Rebellion.[9] The Siamese had ostensibly come to the aid of some tribal groups in order to protect them from these marauding bands by launching a military expedition. This invasion infuriated the leader of the powerful White Tai Deo family who ruled over Lai Chau, an important province on the Sip Song Chau Tai. Deo Van Tri, the *Chau Muong* of Lai Chau, led 600 men of the feared Black Flags and attacked the Laotian royal capitol of Luang Prabang in retaliation for the Siamese action.

The French in turn launched their own expedition into the Tai Highlands and concluded their own agreement with Deo Van Tri. The French "expedition" was headed by the very able administrator Auguste Pavie (hence the name "Pavie Mission"), who had originally arrived in Indochina as a sergeant and stayed on as a civilian colonial administrator after his military service. In order to stop the fighting and build up the area as a buffer state, Pavie offered Deo Van Tri a treaty: if he would renounce his treaty with the Black Flags, Pavie would pressure the Siamese to release Deo Van Tri's four sons from captivity and also offer him French protection (McCoy, 1972). The return of the four sons no doubt was an act that helped ingratiate the White Tai to the French, but no doubt Deo Van Tri also had his own political motivations. Properly manipulated, the technologically superior French would no doubt be a better "ally" than the Black Flags for Deo Van Tri's political aspirations.

For over a hundred years there had been bitter internecine power struggles between the various Deo families for control of the Sip Song Chau Tai. The Lai Chau faction had just won political dominance over the Phong Tho faction through carefully arranged marriages and also agreements with the various bandit groups operating in the area, particularly the infamous Black Flags. It should also be noted that the Tai at time used armed force before the arrival of the French as a means of political control.

Interestingly, the White Tai are the minority tribe in the Tai Highlands. For whatever the reason may be, the White Tai have always controlled or dominated the more numerous Black Tai. With the exception of the White Tai of Phong Tho, it was the White Tai of Lai Chau who would be given nominal control over the Sip Song Chau Tai. In order to stabilize the region politically, the French made a number of physical changes. One curious example is that the provinces of the Sip Song Chau Tai were reorganized and given Vietnamese nomenclature, even though the agreement signed with the French by the Tai provided protection from the lowland Vietnamese. Autonomy per se was granted the Tai Highlands in that their separate treaty was signed with the French and not the Vietnamese.

There were to be more drastic changes initiated by the French. The title of *Chau Muong* was not *Phu Tri*, the Vietnamese word for province chief. This political position was now made elective by the French, which was a move designed to break the power of the Deos as the position of Chau Muong was hereditary within the Deo line. Above the position of Chau Muong all Tai, including the Deo, were excluded. These positions were given to Vietnamese administrators brought in by the French, who were in turn controlled by the French civilian and military administrators. No matter how hard the French tried to break their power, the Deo still had enough prestige to be elected to their previous posts. Aside from the Deo performing a religious function in conjunction with the agriculture, there is also a saying among the Tai: when gold is lost, only gold will replace gold. Deo means gold.

World War II and After: the Militarization of the Tai

This pattern of French rule was to last for about fifty years until the Japanese invasion of Indochina. When the Franco-Vietnamese administration of the Tai Highlands fled to China, the Deo of Lai Chau saw this as another opportunity to take control of the Sip Song Chau Tai. The Lai Chau faction, ever so politically opportunistic, attempted to fill in the power vacuum. To the detriment of the Deo family ambitions, they would be in competition with the Japanese, Viet Minh, and the VNQDD.

When the French returned to the area after the war, they first had to expel the Viet Minh and the VNQDD. Curiously, the Tai Highlands had become a refuge for these two nationalist movements. The VNQDD was using Lai Chau as a base while the Viet Minh had some elements at Son La. The expelling of these groups made the French seem as "liberators" in the eyes of the highlanders (at least some of them). The highlanders had a typical mistrust of the lowlanders' intentions and the VNQDD and the Viet Minh were both viewed as lowlander elements. The Japanese had also left no fond memories behind them, as they had impressed tribesmen into work crews for the airfield at Dien Bien Phu and had not paid the workers (McAlister, 1967).

The Lai Chau Deo were in a favorable position at the end of World War II. For their help in expelling the Viet Minh and others from the region, Deo Van Long, the son of Deo Van Tri, was given the presidency of the Tai Federation when it was provisionally formed by the French. The Tai Federation's boundary matched the boundary of the now defunct Sip Song Chau Tai, which meant that Deo Van Long could still utilize his inter-familial political ties.

With the making of Deo Van Long the president of the Tai Federation, the problems of cohesion became more apparent. Although Deo Van Long had control of his own province, he also needed control of the remaining two provinces of the Tai Federation, Phong Tho and Son La. He branded all the *chau muong* who would not submit to him as "communists" and proceeded to attack them and take them over by force. The French were thoroughly duped by Deo Van Long and continued to give him military assistance and free reign in his "anti-communist" activities.

Deo Van An was the chief of Phong Tho and he was simply bribed into going along with Deo Van Long. It should also be noted that Phong Tho was a strategic location and would obviously bear the brunt of a Viet Minh invasion. Although Deo Van Long was at first against arming "potential enemies," no doubt because he knew better, Deo Van An would eventually receive his own Tai Battalion for his own defense (McAlister, 1967).

Deo Van Long would have more problems in trying to control the province of Son La for the simple reason that the Black Tai controlled this province. The Black Tai uprising against Deo Van Long was led by Cam Van Zung, a Black Tai who had been given arms by the Viet Minh. This revolt was not put down until 1947, but this did not solve anything as Cam Van Zung only went back to the Viet Minh and hid out in the highlands. Bac Cam Qui was installed by Deo Van Long as the new province chief as he was considered more "reliable." His brother had been shot by the Viet Minh, which was obviously a primary factor in his refusal to cooperate with the Viet Minh.[10]

Installing the Bac family to leadership in Son La was no doubt the best political deal that could be made considering the political conditions in Son La. The Bac family constituted one-fourth of the population of Son La, and their stronghold was at Thuan Chau, which was close to Lai Chau. Considering the lack of roads in the highlands, the geographical location of Thuan Chau was ideal for the Lai Chau Deo. The 1st Tai Battalion (1st BT) was raise din this area and consisted of Black Tai from Thuan Chau and Dien Bien Phu (McAlister, 1967:813).

Son La was not like Lai Chau and Phong Tho, in that there was no real political cohesion among the Black Tai as among the White Tai. There was not one dominant family in Son La but five. The Xa of Moc Chau, the Hoang of Yen Chau, the Cam of Mai Son, the Cam of Muong La, and the Bac of Thuan Chau (McAlister, 1967:812).

As was shown politics could not be the solution to the problem of cohesion in the Tai Federation, and the Lai Chau faction would resort more and more to armed persuasion as the means for political control. The opposition families could not in turn be considered "die-hard communists" for going over to the Viet Minh. What they needed was their own military force to counter-balance the Lai Chau military power. The only logical solution was arms, training, and military advisors from the only other military power available, the Viet Minh. As will be shown in Chapter V, they were to receive much more that just military aid.

Why Defend the Tai Highlands at All?

The reason why the French chose Dien Bien Phu as the site of a pitched battle to defend the Tai Highlands is still classified by the French Government. In view of certain facts the site itself was a sensible decision in some respects. The plain of Dien Bien Phu is the largest in the highlands and offered ample room for the French to maneuver men and tanks to break up assault waves, a tactic the Viet Minh were still employing. Initially, the French envisioned the base as having a "firebase" (*aero-terrest* in French parlance) function, which is a large and secure base that patrolling units such as the GCMA's could use for rest and resupply (Fall, 1967:31).

Historically speaking, Dien Bien Phu had been the site of previous confrontations. Under French administration, a junior official was on hand to supervise the local economy. Over two thousand tons of rice a year was harvested in the valley, not to mention one million dollars worth of opium per annum (see Chapter II). Due to the rugged terrain it had been easier and more economical for the French to build hundreds of airfields rather than roads in the Tai Highlands starting in the 1920's. The economy and the airfields would prove to be liabilities during the First Indochina War.

The airstrips in the Tai Highlands, particularly the larger one at Dien Bien Phu, became strategically important with the beginning of World War II. On March 9, 1945, Dien Bien Phu was last-ditch stand by the French against the Japanese, but quickly fell. The Japanese had intended on expanding the already existing airfield as a forward base of operations for attacking Chennault's airbase in Yunnan, China. In the end the Japanese only occupied Dien Bien Phu for two months, but not before developing the hatred of the highlanders. The Japanese had impressed the local populace into work on the airfield and did not pay them (McCoy, 1972).

Shortly after the war, when the French re-occupied the area, Deo Van Long initiated a bitter power struggle with the Lo family of Dien Bien Phu. Through the powers invested in him by the French as the President of the Tai Federation, Deo Van Long was simply able to legally fire the very able Lo Van Hac, a Black Tai and chief of Dien Bien Phu (Muong Thanh). Deo Van Long replaced him with one of his own sons. Lo Van Hac alone could not resist the political and military power of the Lai Chau Deo and so he went and contacted the Viet Minh, who had begun operating in the area. In retaliation for Lo Van Hac's "communism," Deo Van Long imprisoned his wife (McAlister, 1967:826).

Although Lai Chau was still under "French" control in 1953, Dien Bien Phu had been occupied by Viet Minh troops since November 30, 1952. The area was being used by the Viet Minh as a rest and retraining area, with the occupying force being Independent Regiment 148. The bulk of these troops were from areas north of Dien Bien Phu, which would indicate that they were probably Black Tai (Fall, 1967:9-10).

Such was the situation when the French decided to evacuate Lai Chau as indefensible to the (supposedly) more defensible position of Dien Bien Phu. What follows is a discussion of the evacuation preceding the defense of Dien Bien Phu. The events are not only interesting since they illustrate the various problems of commanding tribal units, but also point to the chaotic condition the French had created inadvertently through their political and military errors.

The Evacuation of Lai Chau: Defeat of a Tribal Army

The French commander at Lai Chau, Lt. Colonel Tancart, was ordered on November 13, 1953, to begin the evacuation. Dien Bien Phu had since been secured by a paratroop drop and was being built up into a defensive position by the French. The Tai Partisan Mobile Group

1 (GMPT 1) was the first tribal unit to withdraw under it's commander, Captain Bordier, a Eurasian son-in-law of Deo Van Long.

The evacuation of the remaining troops could only be described as a military and political disaster. Due to a lack of exact records, particularly in the case of guerrilla units, the exact number of troops and casualties will never be known. Although the "official" number of troops is placed at 2,101 Tai and 36 Europeans, there were also hundreds of dependents and other civilians who accompanied the men. Compounding this problem was the desertion rate once the evacuation was announced. Some tribal soldiers decided to stay with their families in the area and take their chances while some were not from the immediate area and simply went home. As one example, with the exception of the indigenous NCO's, the entire 237th Company deserted *en masse* one evening (Fall, 1967:65).

The sheer number of the various tribal units, not to mention the variety of the units themselves, is interesting and deserves some explaining. Unfortunately, exact data is presently lacking, but enough should be available to illustrate how intense the military recruiting by the French must have been:

1) GCMA, or Mixed Airborne Commando Group. The exact number of GCMA units will never be known for reasons cited in Chapter II. It is not known by this author how these units were designated. Sometimes there is a unit number given, i.e., GCMA 8 as at Dien Bien Phu (see Table 5). Sometimes there may be a geographical designation as the "Cardamon" zone between Phong Tho and Binh Lu (Fall, 1967:60,71).
2) CLST, or Tai Irregular Light Companies. For the evacuation of Lai Chau there were anywhere from 20-29 CLST companies present. These companies numbered about 110 men each armed with rifles, submachine guns, and sometimes 60mm mortars. Most were commanded by their own men, but were usually seconded by a French "advisor."
3) GMPT, or Tai Partisan Mobile Group. They are described as wearing French Army fatigues and slouch hats, which would indicate their status more as conventional troops rather than partisans. Upon their (photographed) arrival at Dien Bien Phu, they formed up

and marched in a military formation, which would indicate to this author some degree of training. As another instance of conventional training and the political outlook of these troops, they marched into Dien Bien Phu carrying French and Tai Federation flags. A Vietnamese flag was conspicuously missing (Fall, 1967:20).

4) BT, or Tai Battalion. Eventually there were three Tai Battalions operational in the Tai Highlands. They were well-trained and numbered about 850 men each. Both Black and White Tai were recruited into these units with French and Vietnamese officers being attached as "advisors."

Even though the French had armed and trained thousands of tribesmen in the area of Lai Chau/Dien Bien Phu and organized them into such diverse units, the French never attempted to *politically* modernize the Tai highlands to the same degree they had *militarily* modernized the region. This fact became glaringly evident during the evacuation of Lai Chau. Of the estimated 2,101 tribal soldiers, only 175 along with less than a dozen Europeans survived the march. To this casualty figure can also be added the hundreds of civilians who also disappeared. Aside from a small percentage of deserters, the bulk of which occurred before the evacuation, the overwhelming majority were annihilated by Viet Minh regulars and *local* pro-Viet Minh tribesmen.

The fate of Lieutenant Ulpat is one example of the consequences of the French failure to politically develop even the immediate region of Lai Chau. Lieutenants Ulpat and Guillermit were at first ordered to remain behind and organize GCMA units, but it was discovered that "there was not enough time to learn the terrain and local populace . . . years of neglect . . ." had made this impossible (Fall, 1967:72). It was too late to organize GCMA units, and Lt. Ulpat was then ordered to proceed to Laos where he would link up with the Laotian tribal units and French commandos. Of the six companies under his command, only Lt. Ulpat and twelve other men out of a possible six hundred survived.

The news of the disastrous evacuation must have come as a shock to the tribal Tai allied with the French. Well over 2,000 of their own troops had been annihilated in the span of one week in their own homeland.[11] What is even more incredible is that the French still considered the Tai

Highlands defensible in view of this latest disaster. What follows is a review of the Battle of Dien Bien Phu as it related to the indigenous soldiers and civilian population of the Tai Highlands.

Dien Bien Phu: A Clash of Ethnic Armies

After the arrival of the survivors from Lai Chau, the troops began patrols primarily in order to contact GCMA's in the area. Capt. Guillerminot's reconnaissance had two objectives: 1) contact the Hmong GCMA group at Ban Phathong, and 2) probe Tuan Giao, an important Viet Minh supply point (Fall, 1967:59).

This was not a typical operation from a conventional viewpoint. A few days into the operation, Tourret radioed that he needed not only new maps, but also silver bars. Silver bars were the standard form of payment for highland partisans, particularly the Hmong, and it would seem that his troops felt they needed an immediate raise in pay. The reconnaissance failed to achieve its objective and it was soon realized that Dien Bien Phu could not function as a "mooring point" for various indigenous (tribal) units that would ostensibly control the countryside.

The purpose of this section is not to discuss the military logic of Dien Bien Phu or even the pitfalls of tribal levies. Rather, it is to bring to light the possible pivotal role of tribal and colonial levies in a monumental battle, a fact very often overlooked in any discussion of the Battle of Dien Bien Phu.[12]

As the following tables illustrate, the Viet Minh and French commands present at Dien Bien Phu comprised a high percentage of tribal and colonial units. Not to be left out are the thousands of local Hmong who not only provided the local intelligence, but also provided the guides and human suply transports for the Viet Minh (McCoy, 1972:87).

The battle itself has often been at once referred to and disregarded as a "microcosm" of the First Indochina War (McCalister, 1967:772,832). This author is of the opinion that Dien Bien Phu is a true microcosm of nation-state politico-military practice in regards to the administration of North Vietnam's tribal homelands. In reference to Social Matrix I and the preceding sections of this thesis, France's policies on how to politically develop and help the Tai Highlands led to the inevitable clash and defeat of various tribal politico-military factions at Dien Bien Phu.

CHAPTER IV

Case Studies From South Vietnam

THE CENTRAL HIGHLANDS of South Vietnam would also have its fair share of colonial adventurers and characters not to mention brilliant scholars and far-sighted administrators. Although the list of such individuals is much longer and their accomplishments too numerous to mention, time permits only casual reference. Such names as Alexandre de Rhodes,[13] Sabatier and Mayrena, to mention only a few, may have affected South Vietnam and highland events in the same manner as Auguste Pavie with the Sip Song Chau Tai.

Alexandre de Rhodes was an accomplished linguist and Jesuit priest who arrived in Vietnam in 1627. He was able to perfect the romanized form of writing Vietnamese used to this day known as *quoc ngu*. Formerly written in Chinese characters only the mandarin and elite classes could read and write Vietnamese. The teaching of *quoc ngu* would only weaken the mandarins' power over the populace and was therefor viewed by them as a threat to the Vietnamese political structure and their positions within it (Karnow, 1982:59).

The exploits of such individuals as Mayrena and Sabatier directly affected the Montagnards, or "mountaineers," of the highlands of South Vietnam. The Montagnards of the South Central Highlands of South Vietnam have an estimated population of circa one million (see Table 9). Numbering over twenty tribes, the Jarai and Rhade are the largest and combined, the two constitute roughly one-fourth of the total Montagnard population. What follows in the next two sections is a review in more detail of the exploits and accomplishments of these two men.

Mayrena and the Catholic Mission in Kontum: A Heart of Darkness?

One of the first expeditions into the Central Highlands of South Vietnam with the purpose of forming a French-Montagnard alliance was undertaken in 1888. The man in charge of the expedition was Charles David, better known by his *nom de guerre* of Mayrena. David was a former professional soldier who had served in Cochin China in 1862 and with the French during the Franco-Prussian War of 1870-71. He was accompanied by a friend, Alphonse Mercural, described as "a former soldier with a sordid reputation" (Hickey, 1982).

Due to a number of serious revolts and uprisings in Southeast Asia during this period, Mayrena had proposed to the French authorities that he and Mercural mount an expedition in order to investigate rumors that the Montagnards were being recruited into the uprisings taking place along the Cambodian-Laotian-Vietnamese borders. It was suggested by the French that they contact Pim, a Bahnar chief on favorable terms with the French. It was hoped by the French authorities that a confederation of the Bahnar, Rengao and Jarai Hodrung tribes could be established with Mayrena as their "leader." This would enable the French to have a secure region in Kontum Province to protect against further uprisings of foreign incursions.

Just prior to the arrival of Mayrena, Father Guerlach of the Catholic Kontum Mission had complained of attacks by the Jarai and Sedang tribes not only on the Mission itself, but also on the Bahnar and Rengao tribes he was seeking to convert. Father Guerlach persuaded the Bahnar and Rengao chiefs to send their warriors with him on a military action to defeat the Jarai and Sedang, but due to the fact that the Bahnar and Rengao had never defeated these tribes in a war they were skeptical of Guerlach's promises. Nonetheless, 1200 tribesmen showed up for the campaign and in February 1888 they quickly defeated the Jarai with Father Guerlach personally leading the attack. During this time Pim, the Bahnar chief, was able to conclude his treaty among the other chiefs for the Bahnar-Rengao Confederation (Hickey, 1982b).

Mayrena arrived in the area around April 1888 and with the backing of Guerlach proceeded to expand the Bahnar-Rengao Confederation into a larger "kingdom." After concluding a number of treaties, Mayrena declared the founding of the Kingdom of Sedang on June 3, 1888. On the 4th of July an alliance was made between the Bahnar-Rengao Confederation and the Kingdom of Sedang. Mayrena crowned himself as King Marie I and Mercural became the Marquis of Henui.

Although this episode might be dismissed as a ludicrous historical footnote, one should not dismiss the various aspects of culture contact. Not only were the inter-tribal alliances formed for mutual defense, but the Bahnar and Rengao were able for the first time to defeat the powerful Jarai. Mayrena was then able to form an even larger, albeit loose, coalition from the tribes they had just defeated. One wonders what the end result of this larger coalition would have been if Mayrena would have remained in the highlands. He left in late 1888 to procure modern arms for his new "army" and even left an order with a Chinese tailor in Haiphong to have one thousand uniforms made (Hickay, 1982a:233). After a year in Europe raising money, he left for Vietnam with a large shipment of modern arms. The French authorities were alarmed at such independent developments and Mayrena was refused permission to land in Indochina, along with having his weapons confiscated. He died near Singapore in November 1890 never having realized his dreams.

Pim alone was not able to establish a lasting Bahnar-Jarai-Rengao Confederation. He remained the leader of the Bahnar-Jarai Confederation until the French stepped in an dissolved the Confederation in 1895. Although Pim had been stripped of his powers by the French, the lessons he had learned were not forgotten. Nay Moul, a Jarai leader that will be described later in this chapter, was a protege of Pim.

Pacification: Consolidating the Highland Minorities

The French decided in June 1895 that there existed a legal basis for actually establishing a presence (i.e. French administration) in the Central Highlands. The French had since realized to some degree that the economic "El Dorado" that de Rhodes had described did not in fact exist (Karnow, 1983:60), but the French were later to justify their colonization of Vietnam as a *mission civilitrice*, or the import of French civilization to an otherwise "barbarian" country (Karnow, 1983:79). In order to import such lofty idealism the French obviously first had to secure the region. It was decided that the best manner to implement authority was by establishing military-administrative posts in order to maintain regional control (Hickey, 1982a:272).

In the area of Kontum, the Kontum Mission was again put in charge of this latest scheme. What is interesting to note is that the Mission was given the authority to arm Montagnards at their discretion for defense. Tournier, the Commandant Superior of Lower Laos, offered the Kontum Mission 35 carbines and 200 rounds of ammunition. Although this may seem like a small amount of weaponry, one should compare this to the American CIDG program of the 1960's (see below). The number of villagers trained as defenders would outnumber the weapons provided, as the villager on duty would turn over his weapon to the next man when his "shift" was over. This system of defense is particularly effective when used in conjunction with a back-up "reaction force" of full-time professional soldiers (see "Strike Force" below).

Leopold Sabatier, one of Frence's more able administrators in Vietnam, had also been interested in arming villagers. A *Garde Indigene* had previously been established in Annam (South Vietnam) in 1886 (Hickey, 1982a:273). The theory behind the *Garde* was that villagers would better accept "pacification" from their own kind rather than "foreign" soldiers. One incident which illustrates this need occurred in 1905 when the French decided to integrate the territory of the Sedang, Rengao, Bahnar and Jarai into the Plei-Ku-Der (Pleiku) Province. The French mistakenly sent a French-commanded unit of lowland Vietnamese into the area which touched off a series of uprisings by the Montagnards (Hickey, 1982a:286).

Pasquier, the Resident Superior of Annam, was impressed by the accomplishments and innovations of Sabatier. Of particular interest were the Franco-Rhade School and the *Garde Indigene Moi* ("*moi*" means "savage" in Vietnamese). Pasquier felt is would be better if the various Montagnard units would also be commanded by men of their own ethnic group. He also felt that these militiamen, when their time was up and they returned to their respective villages, would make excellent cultural brokers and were " . . . the best means for propaganda for the French administration" (Hickey, 1982a:303). The recruiting of Montagnards was to intensify and the *Garde* in 1930 would include 31 Rhade (Hickey, 1982a:334).

In 1931 the Battalion Montagnard du Sud Annam (Highlander Battalion of South Annam) was formed. Outfitted as a conventional unit, they would not see action until 1940. Due to the Japanese invasion of China, France saw it necessary to build up its' forces in Indochina and recruitment was intensified. The first action the battalion saw was not against the Japanese but against the Thais. The Autonomous Brigade of Annam, of which the Highlander Battalion was a part, was involved in this border war.

A number if interesting events were to involve the Montagnards serving in the French forces from 1940 to 1946, the year the First Indochina War began. Soon after being involved in the Thai border war, Paris was taken by the Germans. This in effect left the French in Indochina to fend for themselves as best they could. As a result, some French officers and administrators collaborated with the Japanese, while others threw in their lot with the Free French (via Force 136).

By March 1945 the Japanese ordered that all non-Japanese military units in Vietnam be disarmed and report to the Japanese due to the actions of Force 136. Those who could escape did, but a number of Montagnards in the army ended up in labor camps, such as Touprong Ya Ba and Y Pem Knoul (both Rhade). One Rhade man who did not run away but stayed and was made a battalion commander in a Japanese-Rhade unit was Y Thic Mlo Duon Du.[14]

A strange odyssey was to befall the Montagnards in part of the 4eme Bataillon Montagnard du Sud Annam. It is related how "most of the personnel in the battalion were Jarai and Rhade, with some Sedang and Bahnar" (Hickey, 1982a:377). When the Japanese ordered the units to disarm, most of the Montagnards in the 4eme Bataillon were on operations in the Red River Delta area of North Vietnam. The French officers instead ordered the Montagnards to disperse into the forests rather than surrender. They were to regroup at a designated area and then to escape to China where they eventually remained for about five months.

After a number of French soldiers and French Foreign Legionnaires arrived in China from Indochina, the Montagnards and French soldiers were re-outfitted into a new unit called the Groupement Quiliguini. By the time Groupement Quiliquini reach Ban Me Thout in July 1946, World War II was over and the First Indochina War had begun. R'com Hin and twenty-three other Montagnard officers and NCO's of Groupement Quiliquini were reassigned to the 3eme Battalion, 22eme Regiment de l'Infantrie Coloniale stationed at Bien Hoa.

Other Montagnards formerly in the military asked to return to active duty. Y Sok Eban was reaccepted and reassigned to camp at Poste du Lac. Y Pem Knuol was reassigned to the 5th Far East Infantry Battalion, which was composed of Montagnard volunteers (Hickey, 1982a:390). Y Bih Aleo and Y Tuic Mlo Duon Du, formerly of the Viet Minh, were allowed to rejoin, but not until after Y Tuic served six months in jail. The two brothers Touprong Hiou and Touprong Ya Ba were also jailed for Viet Minh activities, but Ya Ba was later to be given the rank of sergeant in a local Montagnard militia.

The First Indochina War

As stated earlier, a variety of Montagnard units and mixed units were operational by the beginning of the war in 1946. Ho Chi Minh and General Giap soon shifted the emphasis of the war to the highland regions of North Vietnam (including southern Laos) and the Central Highlands of South Vietnam. In 1950 the situation in the Central Highlands had deteriorated to the point that it was referred to by the French as *pourissement* (rotting away).

The French decided to again expand the Montagnard units which resulted in March 1951 in the formation of the 4th Vietnamese Light Infantry Division,

> " . . . better known as the 'Division Montagnard' because most of it's personnel were from such upland ethnic groups as the Rhade, Jarai, Bahnar, Mnong, Sedang, Maa, and Hre" (Hickey, 1982a).

Three battalions were composed exclusively of Montagnards, the 1st, 2nd, and 7th Battalions, with one mixed battalion of Vietnamese and

Montagnards. Although divisional officers were French and Vietnamese, a new officer's training center, the Ecole Militaire Regionale du Lac, had just graduated seventy-five Montagnards *aspirants* with another fifty soon to graduate. A partial roster of two of the battalions of the Montagnard Division was provided by one author:

> "The 5th Battalion had among its personnel many Rhade and Jarai, including Nay Moul, a sous-lieutenant who had been one of the first Jarai to join the French Army; Rcom Hin; and Rcom Pioi, a former school teacher from Cheo Reo whose brother Rcom Briu had joined the Viet Minh. Other Jarai in the battalion were Nay Honh, a graduate of the Academie Militaire; Rcom Ko; Nay Dai; and Siu Nay. The Rhade included Y Jao Nie Buon Rit, who was married to Rcom H'bu, the eldest daughter of the famous Jarai chief Nay Nui . . . Other Rhade were Y He Buon Ya, who also was married to a Jarai woman and lived in Cheo Reo, and Y Pem Knoul.
>
> " . . . The 7th Battalion, whose symbol was a highlander machete and an elephant's head, was composed for the most part of men from the ethnic groups in Haut Donnai province. According to Touneh Han Tho, most were Chru, Sre, and Maa. The commander was Touprong Ya Ba, brother of Touprong Hiou . . . Other officers Touneh Han Tin, half-brother of Touneh Han Tho, and two Sre aspirants, K'sau and K'nam. Among the noncommissioned officers were Touneh Ton and K'tu, a Sre" (Hickey, 1982a:407-9).

In late 1953 and early 1954, just prior to the fall of Dien Bien Phu, more Montagnards would be recruited into a variety of newly formed units: Mobile Groups 41, 42, and 100. Of Mobile Group 100, the 1st, 2nd, and 3rd battalions consisted primarily of Rhade, Jarai, Bahnar, Sedang, and Chil tribesmen recruited after Mobile Group 100 reached the highlands (Hickey, 1982a:430). Mobile Group 100 was involved in one of the last actions of the war when it was ambushed and almost wiped out.

Not to be left unmolested would be the irregular units of Montagnards. The French had built up GCMA units in South Vietnam as in North Vietnam, and one other organization that was somewhat akin to the GCMA was the Detachment Legers de Brousse (Light Brush Detachments), also composed of Montagnards (Hickey, 1982a:409). The French had also established a myriad of diverse Montagnard units such as guard units, village self defense units, and even inter-village supply and communications units.

What follows in the next sections is a recounting of the American attempts at counterinsurgency. The French had realized after almost ten years of warfare that conventional units, tribal or otherwise, were not the answer to the First Indochina War. But such programs as the arming of tribal irregular troops in a GCMA style were begun too late to be effective. As explained in Chapter II, the American experts at first refused to continue any French programs. As a result of this, the American programs in practice during the Second Indochina War would unintentionally mirror the French efforts.

Second Indochina War: Tribal Mobilization (Again)

Starting in the early 1960's, the American effort would deem the winning over of the Montagnards as essential. A number of factors led to this opinion:

1) The "Ho Chi Minh Trail" went from North Vietnam through Laos and entered South Vietnam through the Central Highlands.
2) The insurgents were intimidating the villagers, which necessitated the relocation of unprotected villages into strategic hamlets.

President John Kennedy was especially impressed with the counterinsurgency concept of warfare. During his visit to the Special Warfare School at Fort Bragg, North Carolina, he pledged his personal support to the "new" form of warfare. As a personal touch, President Kennedy awarded the Special Forces the official right to wear the green beret, the most obvious symbol of their elite status within the military.

Presidential support would also enable the Special Forces to perform a very specific and specialized mission in Vietnam without interference by regular army units and their commanders. This mission, the training, arming, and leading of indigenous units will be discussed in the following section. What is important to note is the number of tribesmen trained and armed by the Americans in comparison to the French.

The CIDG Program, cited earlier in Chapter II, was a program designed for villagers to have a mutual defense of each other. In a word, the concept was purely defensive in nature. Concurrently, aside from village soldiers trained solely for defense, there were also units designed for offensive military operations. To the 1962 figure of village defenders (hamlet militia) can also be added 300 border surveillance troops ("trail watchers"), 2,700 mountain scouts, and about 5,300 Popular Forces (see Table 11). The grand total of indigenous forces is at least 33,300 irregular forces. By June 1963, at least 20,000 more Strike Force and hamlet militia would be trained (Kelly, 1972:37). But due to the intensification of the war, the character of the CIDG Program as a whole would change:

> "The higher priority given the border surveillance mission after July 1963 caused a shift of the principle CIDG effort from the interior to border sites and was the reason for the turnover or closeout of seventeen of the eighteen camps relinquished between August 1963 and March 1964. Inadequate initial area assessment led in some cases to the selection of unproductive sites that later had to be relocated. Some camps were situated on indefensible terrain or had limited CIDG potential. Other camps were moved or closed out altogether because of the *discontent of the strike force, which had been recruited from a distant area because of the lack of local resources*" (Kelly, 1972:44, italics added).

The initial program of arming tribesmen for their own defense had obviously grown out of control by 1963. The arming of too many tribesmen too fast had suddenly become a liability to the war effort. The American programs were originally designed to provide tribal areas with their own defense composed of soldiers that reflected an area's ethnic composition. As Table 11 illustrates, tribal soldiers were now being sent to defend territory foreign to them. What the following section will briefly discuss is the new emphasis placed by the American military experts on the conventional tribal soldiers.

CIDG After 1963: The Conventionalization of Tribal Troops (Again)

The conventionalization of indigenous forces in Vietnam could be attributed to a number of factors. Specifically, the border surveillance was not as effective as it should have been. In order for the camps to have been effective at sealing off the borders, they would need to be spaced at intervals of twenty miles apart. This figure was based on the fact that each camp could adequately patrol only within a twenty mile radius. But the highest density achieved was only twenty-eight miles between camps (Kelly, 1972:52).

What this problem necessitated was a more radical approach in an attempt to seal the borders. Almost the entire emphasis would be placed on expanding the Strike Force troops. It was hoped in November 1963 that 20,000 Strike Force troops could be trained by July 1964. In March 1964, the CIDG would become conventionalized following the example of Table 10. Except for very minor changes (i.e. company strength) the CIDG guerrilla company of 1964 is surprisingly similar to the French CLST's of 1954 (see above).

The emphasis was not to be placed on the previously para-military forces within the CIDG. Very few hamlet militia would be trained as this was now considered secondary. This is not to say that Montagnards would not still become more conventional in nature. As Table 11 illustrates, the ethnic composition of the Strike Force troops, particularly of the camps in the highlands, was still predominantly of Montagnards.

Not surprisingly, the reader will note a prominence of certain tribal groups within the Strike Force: the Rhade, Jarai, Bahnar, Sedang, Mnong, etc. The reader may have noticed that the groups represented on Table 11 are not representative of the circa twenty-nine Montagnard groups in South Vietnam, but they are representative of the tribal groups militarized by the French during the First Indochina War.[15]

The South Vietnaese Governemt was alarmed at the number of Montagnards in various military units, in particular the Strike Force.

There had been a history of unrest in the highlands and due to the fear of a revolt, this time armed with modern weapons, the Government ordered the number of weapons in the highlands to be reduced by 4,000. This action was to lead to a large and well organized uprising of the Montagnards usually referred to as the FULRO Revolt.

The 1964 Revolt in the Highlands

Prior to the FULRO Revolt a number of events, aside from the arming of tribesmen, had alarmed the South Vietnamese Government. The Montagnards had begun to organize politically since the late 1950's (see below) and were becoming more militant. It was not merely the arming of tribesmen, but the availability of weapons in conjunction with independently politically organized tribal groups that had alarmed the lowland Vietnamese.

The Montagnards had enjoyed a certain amount of local autonomy under the French. The French had also protected the Montagnards from encroachments by their traditional enemy, the lowland Vietnamese. The departure of the French administrators in exchange for lowland Vietnamese administrators no doubt caused alarm among the Montagnards. Soon after the departure of the French political support, the Montagnards began forming political organizations in order to voice their grievances to the South Vietnamese Government. The first organization was known as the Front pour la Liberation des Montagnards (or FLM), followed by the broader organization known as the Bajaraka (see chapter V).

During the 1960's, a number of Montagnard delegations from the Bajaraka had approched the lowland Vietnamese generals in charge of their provinces. Of particular interest is the continued requesting by the Bajaraka of the *Statute Particulier* of Emperor Bao Dai, which had been proposed in May 1951 by the French but ignored by the Vietnamese. What this decree called for was the "free evolution" of the Montagnards, with the government to provide them with socio-economic development programs (Hickey, 1982b:95).

On 25 August, 1964, the Saigon government sponsored a meeting of Montagnard leaders in Dalat. The leaders had also been instructed to bring along a list of their "aspirations." Aside from requesting more health and educational programs, the delegation again requested that the *Statute Particulier* be implemented.

On 20 September, less than one month after the meeting, Montagnard troops trained by the Americans revolted in five Special Forces camps: Buon Sar Pa, Bu Prang, Ban Don, Buon Mi Ga, and Buon Brieng. Showing surprising organizational skills, Montagnard leaders had organized the armed revolt of more than 3,000 Montagnard soldiers to occur spontaneously over a wide geographical area. On the day of the revolt, the Montagnards announced the founding of their new nationalist movement, Front Unifie de Lutte des Races Opprimees (The United Struggle Front of Oppressed Races, FULRO).

In a dramatic and violent show of force, the Montagnards again presented a list of demands to the South Vietnamese Government. Aside from again demanding that the *Statute Particulier* be institute, they also demanded more schools and medical aid be provided (Johnston, 1966:128). The South Vietnamese were dissuaded from attacking the FULRO troops and instead a negotiated settlement was arranged by the Americans. Many of the Montagnard demands were met by the South Vietnamese with the exception of autonomy. In exchange for accepting a negotiated settlement, the Montagnard soldiers that took part in the revolt were granted amnesty.

Amnesty or no, many of the FULRO leaders left for sanctuary in Mondulkiri, Cambodia. With the intention of carrying on their struggle, the FULRO leaders left for Cambodia with anywhere from 500-2,000 Strike Force soldiers. This was by no means the end of FULRO or of rebellions by American-trained Montagnard soldiers. FULRO troops continued to attack both North and South Vietnamese units throughout the duration of the second Indochina War.

A further discussion of the possible reasons for the evolution of a radical ethnonationalist movement among tribal soldiers is reserved for the following chapter. Let it suffice to say at this point that the French effort in the Tai Highlands resulted in an intertribal war, whereas the American effort resulted in a previously unknown tribal cohesion and large-scale revolt.

Summary of Chapter III and IV

In practice, the two different doctrines of unconventional warfare operated on two social planes. Although the initial contact of tribal groups was with their indigenous leaders, the actual targets for converts to the respective political causes were different. According to one American analyst:

> " . . . our efforts have been directed at those segments most influential in a modern society, e.g., politicians, government officials, labor and business leaders" (Johnston, 1966:104).

According to General Giap, the communists saw a practical purpose in controlling the lower levels of society:

> "People in towns have chairs, tables, shoes, beds—you can't eat those things. Country folk have rice, eggs, chickens, pigs. Remember, those who rule the countryside rule the country" (Dommen, 1971:69).

In correlation to the unconventional warfare formulae presented earlier, the co-efficient of the variable "GW" would be the Tai, Tho, Montagnard, etc., soldiers recruited and trained by the French, North Vietnamese, and Americans respectively. As was evident in the two preceding chapters, tribal soldiers proved to be quite flexible and could be trained to fight as true guerrillas and as conventional troops.

Although the tribal soldier seemed to perform best in the unconventional mode, he was an invaluable asset to the nation-state. He knew the people, language, and terrain of specific regions, thereby providing the crucial link as cultural broker between the nation-state and tribal population in general. The questions that will be considered in the conclusion (Chapter V) are whether the traditional patterns of leadership are undergoing change during protracted warfare.

CHAPTER V

An Anthropological Interpretation of Change

WHAT SHOULD BE evident from the preceding chapters is the dependency of both systems of communist insurgency and Western counterinsurgency in exploiting the fears and animosities of the various tribal minorities. As was historically illustrated in Chapters III and IV, contact by a nation state third party inevitably led to the "necessity" to stabilize the region. The inevitable solution was to militarily modernize the dominant tribal group(s) as a means for stabilization. Paradoxically, this would actually destabilize a region as the smaller tribal group(s) or tribal political faction(s) would in turn search for the same military assistance form a rival nation-state in order to achieve parity.

Unfortunately for the tribal groups contacted by either third party they were never fully aware, by design or nature, of their pivotal role in international politics. Induced to join a "cause," tribal leaders could be motivated for a variety of reasons: promises of autonomy, economic advantages, or even simple revenge to name just a few, as evidenced by the historical reviews. As the war in the highland regions intensified, with more tribal levies being raised from varied tribal groups, the war increasingly took on the characteristics of an inter-tribal conflict.

As the conclusion will illustrate, the Western methods of counterinsurgency generally did not entail much appreciable change in traditional leadership. Most important to note concerning the communist insurgency method is that the traditional family/clan oriented society through the *parallel hierarchies* and the front organizations.

North Vietnam After the First Indochina War

The ethnic groups in the Tai Federation, as explained in Chapter III, had chosen Barth's third option as a desired course of action. The question for the various tribal leaders was with whom they would ally themselves for the fight for autonomy. They could either ally themselves with the French through Deo Van Long, a White Tai, or with the Viet Minh through Lo Van Hac, a Black Tai. Among the Tai under the indigenous system, it was the *chau muong* who retained the right to raise tribal levies (see Table 13). During the First Indochina War it is also these same individuals, whether pro-French or pro-Viet Minh, who would retain their former leadership positions.

Table 7 illustrates how the war in North Vietnam was primarily an inter-tribal war. True to the warfare theories of Ho and Trinquier, the war in tribal areas according to the diagram also illustrates how tribal rivalries were exploited to the fullest. As evidenced by the historical data provided in Chapters III and IV, inter-tribal warfare had existed before the French had arrived. In the view of this author, the data supports the fact that the French did not instigate inter-tribal warfare as a by-product of colonialism. What is important to note is that the French were the first nation-state to offer modern arms into an already existing volatile situation. The logical conclusion to modern militarization was an ever escalating conflagration that involved virtually every tribal group in the North.

Ho was to waste no time in keeping his promise on the issue of autonomy to the tribal leaders who had supported him in the First Indochina War. During the Battle of Dien Bien Phu, a Congress of Minorities was held with 140 highland leaders from twenty different ethnic groups. The chairman of the meeting was Chu Van Tan and the purpose of the meeting was ostensibly to discuss the future autonomy of the homelands (Fall, 1963:149). After the French surrender, a number of "autonomous" zones were established in North Vietnam: the Tai-Meo Autonomous Zone, the Viet Bac Zone, and the Lao-Ha-Yen Zone.

The Tai-Meo Autonomous Zone, formerly the Tai Federation, was established on May 7, 1955, and four days later elected a congress with Lo Van Hac as president. The congressional representatives included ten Tai, five Hmong, two Vietnamese, two Muong, one Man, and five representatives of other minorities. There were also two Hmong zones called Thua Chua and Mu Cang Chai. At first, the original sixteen *muongs* of the of the defunct Tai Federation and Sip Song Chau Tai were recognized with the exception that instead of Lai Chau or Phong Tho, Ban Chieng Ly (Tuan Chau) became the new capital. The entire zone was fully integrated into North Vietnam on October 28, 1962, as the Tay Bac Zone, which means "Northwest." The sixteen *muong* were also changed back to the three provinces of Lai Chau, Son La, and Nghia Lo (Fall, 1963:149).

The Tho became autonomous on August 10, 1955. Chu Van Tan was elected as president of the Viet Bac Zone, a position that is similar to the historical Nan Chao Kingdon of China mentioned in Chapter III. The capital was Thai Nguyen, which had been Ho Chi Minh's headquarters during the First Indochina War. The Lao-Ha-Yen Autonomous Zone was established on March 25, 1957, but became ungovernable due to the complex ethnic composition. It was abolished two years later and assimilated into other zones.

Although these zones were integrated into the Democratic Republic of Vietnam, there was still the Committee on Nationalities, of the *Uy Ban Dan Toc*. This office is supposedly to be consulted on all matters pertaining to minorities. The chairman of the Committee on Nationalities in 1966 was the former general of the 312th Division, Le Quang Ba.

The question is how did the zones become assimilated into the larger nation-state? The tribal groups had not only expressed their desire to follow Barth's third option, but had also been promised autonomy by the Viet Minh. The government of the zones at the time of "independence" was nothing more than the UBHC's discussed in Chapter I. As will be remembered, UBHC's had been imposed by the Viet Minh in order to give the various tribal factions unity of command up to the Central Committee under Ho. How the UBHC's became attached to the indigenous political systems, thereby paving the way for assimilation, will be discussed in the following section.

Parallel Hierarchies: the Assimilation of Tribal Leadership

Table 13 will facilitate in explaining how politico-military power is slowly usurped from n indigenous political system by the UBHC's. The graph coordinates prominent indigenous politico-military offices with the prominent corresponding offices of the parallel hierarchy. A more detailed example of the imposition of such a system in relation to the Tho is presented below in conjunction with Table 12.

Once a parallel hierarchy has adapted itself to the Tho system it is obvious what offices will eventually become absorbed by the Viet Minh UBHC. For example, a UBKC/HC will be imposed at the *canton* (*tong*) level. The *chau tong* will now have to report to his own UBKC/HC counterpart, the president, for all decisions in order to make sure that they conform to Party lines. The *pho tong* will also have to clear his duties with the UBKC/HC secretary. But most importantly, the military role will be taken out of the hands of the indigenous system. The *chau doan*, *linh*, and *doi le* will either cease to exist, or become figureheads as all military decisions will also be taken over by the UBKC/HC. The reader will remember from chapter I taht the UBKC/HC is actually a military organization by nature and not solely administrative.

The Tho may offer different problems for analysis than other tribal groups such as the Hmong or the Tai (see Table 13). As stated earlier, the Tho had begun a process of "Vietnamization" since the 16th Century with the arrival of the Vietnamese Mandarins and the emergence of the Tho-ti. Other changes were to take place in the Tho culture. Under Vietnamese influence the kinship system of the Tho was to change from bilateral to patrilineal. Tho religious practices would also change, with the advent of an ancestor cult and family land inheritance, all of which may be due to the import of Confucianism (Schrock, 1972). But these would not distract from the major concern of this thesis, which is the political impotency of indigenous officials due to the effects of the parallel hierarchies.

Assimilation solely through the offices of a parallel hierarchy would be impossible were it not for the concurrent and interrelated actions of the front organizations. The function of the front organizations should not be overlooked within the framework of the parallel hierarchy as the primary vehicle for social change (assimilation). What follows in the next section is a review of the land reform policies of the China and North Vietnam and the true function of the front organizations.

Front Organizations and Land Reform: Assimilation of the Masses

In China and North Vietnam, the true purpose of collectivization and land reform was not revealed to the "masses" until after the war. The various slogans such as "land to the tillers" was a powerful incentive to the largely landless peasantry. Promises of local autonomy were also heady political stuff for the tribal groups that had voiced a desire for autonomy (Barth's third option). The communists needed first to isolate the minorities and then to motivate and mobilize them for military action. Isolation from the legitimate government was achieved through the parallel hierarchies with mobilization for "autonomy" being achieved through the various front organizations.

Both Zasloff (1973) and Dommen (1971) provide detailed accounts of the front organizations in their books. Front organizations would be composed of such groups as "Youth Associations," "Peasant Associations," "Young Girls Associations," etc. (see Tables 1 and 15). These organizations, either by design or nature, will eventually break down the traditional family/clan social structure into various committees. A son, for example, would no longer be able to consult with his father, or any other family member for that matter, on personal decisions, i.e. marriage, travel, occupation, etc. He would now have to propose his decisions to his respective front organization members for their approval.

Maintaining control and discipline in the village would be a very easy process for the parallel hierarchy cadre once the front organizations are in place. A village chief could be discredited and forced to resign through a "slander campaign" that had been initiated from one or more of the front organizations through cadre supervision. Through the prodding of cadre officials a village chief could be accused by fellow front organization members of such crimes as "low food productivity." The village chief would then be forced to read a prepared confession to the assembled villagers,[16] an act which was utilized by the cadre to produce an "extreme loss of face and public humiliation" (Hosmer, 1970:60). The political power of such an individual within the community would thereby be effectively neutralized and the manipulation of the village affairs by cadre officials easier.

As explained in Chapter I, the cadre could call on not only the secret police, but also the *Dich Van* for assistance in certain situations. Certain "crimes" could be considered very serious and could warrant a transfer to either a thought reform camp or even execution. Particularly serious crimes were those that concerned the more recalcitrant villagers. Such individuals were categorized as "hostile" or "backward" civilians "committed to their social class" (Hosmer, 1970:65). One youth, for example, was sentenced to a camp for the seemingly innocuous crime of "throwing sticks and stones at one of our meetings" (Hosmer, 1970:67).

The activities and function of the front organizations would take on a new dimension once the war had been won in China and North Vietnam. The end of the war in China heralded the beginning of a campaign of revenge of "People's Tribunals," a campaign designed to **physically** rid the peasantry and tribesmen of their own leadership. It was the front organization cadre that whipped up the members of the various front organizations into the:

> " . . . yelling mobs, the hysterical denunciations, and the doing to death of some million unarmed people, perhaps for more—a process of fanaticizing the masses and *destroying any alternative rural leadership* in order to prepare the way for what the Maoists had from the beginning been aiming at: the utter subjugation of the masses in order that the small elite Party should be able without hindrance to impose its orthodoxy upon this very great people" (Fairbairn, 1974:120-21, italics added).

It will be remembered from Chapter I that the success of Mao and his military theories rested upon the mobilization of the peasants as soldiers. The peasants realized too late that they had been duped and their own aspirations for autonomy counted for little:

> " . . . the peasant armies had fought, under altogether phony propaganda slogans, in order to encompass their own destruction. They became an 'army of the people' in order to

> destroy themselves a s they had wanted to be: free peasants. They had fought out of local desperation, in order to bring into being the most totalitarian society the world has ever known" (Fairbairn, 1974:121).

This same pattern of events from China would repeat themselves in North Vietnam in 1956, particularly in the highlands regions. Estimates vary at the number of executions, but this is academic when one considers the social class of people targeted for execution. Gerald Tongas, an eyewitness to the event, states:

> " . . . there is strong reason to believe that the campaign was also used as a convenient pretext for *purging the North of potential oppositionists* and for those who had worked for or were suspected of having cooperated with the French" (Hosmer, 1970:97, italics added).

Those individuals who would be purged from the general population of the Northern tribal groups fell under two categories: 1) those who had fought with the French in such units as the GCMA's (see Chapter II), and 2) virtually any tribal leader or potential leader who could muster more prestige than the local cadre/Party official (see Tables 12 and 15). It could be argued whether or not the true purpose of the communist leadership was simply political control rather than assimilation. Whatever the case may be, one fact does remain clear from this analysis of communist unconventional warfare as practiced in tribal regions: the imposition of parallel hierarchies, front organizations, and land reform will lead to a group's eventual assimilation (Barth's second option).

Leadership and Western Counterinsurgency in South Vietnam

There are literally over a hundred ethnic groups world-wide that have been contacted by Western nation=-states for military service since World War II. On the basis of publically available ethnographic material (i.e. personal data, kinship charts, etc.) only in the instance of

South Vietnam was there enough important relevant material in order to initiate a probe into the possible effects of militarization concerning indigenous leadership.

A list of names was compiled by Dr. Hickey of what he considered to be the one hundred most prominent Montagnard leaders in South Vietnam. The original purpose of his list, to be used in conjunction with his kinship charts, was to illustrate through intermarriage the existence of a "new elite." The purpose of this author's own re-analysis is to illustrate the continuation of traditional leadership patterns during two decades of military and civil service to the nation-state.

Of particular interest is the retention by the traditionally powerful clans of their political power during accelerated warfare. For example, in the case of the Jarai, certain clans are noted for producing prominent leaders, such as the Ksor and Nay clans. A male also gains prestige by marrying into the Rcom clan. Referring to Table 17, the reader will note that the majority of the Jarai leaders are still of the Ksor (20%), Nay (25%), and the Rcom (25%) clans. This pattern is maintained in the case of other Montagnard gropus. In the case of the Chru, traditional leadership is among the Touneh, Touprong, Jalong, Banahria, Sahau, and Klong Drong clans (Hickey, 1982a).

Just because a Jarai or Chru decides to move out of the village into nation-state military or civil service does not necessarily make him a possible political usurper for village chief. Granted the full-time tribal soldier will be able to collect assorted new prestige items (sunglasses, radios, motorcycles, etc.) which he will be able to return to his village through the process of redistribution. This would no doubt add to his prestige within the village, but leadership is dependent on a variety of factors. The odds of achieving leadership among the Montagnards solely through nation-state service without the other factors being taken into consideration seem very slim as evidenced by Table 17.

Possibly the most important element in leadership determination among the many Montagnard tribes is the manifestation of "charisma" and "strength." Among the Jarai both are known repectively as *kdruh* and *kotang*. Nay Moul, the protegee of Pim, was considered by Jarai shamans to have had both characteristics early in his life, years before he joined the French army (Hickey, 1982a:335). A similar belief exists among other

Montagnard tribes, charisma being called *dhut* in Rhade and *gonuh seri* in Chru. These characteristics manifest themselves early in life and a child is predestined to lead an extraordinary life. Military service can then be considered a "confirmation" of what is already "known" by the tribe: an individual's leadership ability and courage.

The fact that there was no appreciable leadership change among the Montagnards as opposed to the highlanders of North Vietnam under the communists may be due to the nature of the war. The two unconventional doctrines as espoused by communist and Western theoreticians was shown to be diametrically opposed. The reader will remember the mathematical formulae RW = GW + PW and COIN = GW. These formulae were expanded upon at the conclusion of Chapter IV to illustrate that two different social strata were targeted for recruitment. In the Western approach contact was initiated and **maintained** with the political factions then in power in a specific region. The village chief recommended the men whom he considered loyal to his benefactors (see Chapter II). His power was not circumvented by any political organization remotely resembling a parallel hierarchy.

An absence of front organizations and a parallel hierarchy would mean the maintaining of the status quo in a village. Decisions deemed important by the family, clan, etc., would be discussed in their proper setting without outside interference from political cadre. The village social structure as illustrated in Table 14 would remain virtually unchanged. For better or worse, the leader of a village or political faction would remain in control unchallenged. Not to be overlooked is the hard fact that the village/political leader in question would have his position enhanced by the acquisition of modern weaponry. The question to be raised is what will the leader do with his new weapons? Would the arming of one faction lead to political disequilibrium in the region?

Crime and Counterinsurgency: The Abuse and Corruption of Power

A somewhat bizarre phenomenon would occur with the arming of tribal soldiers, particularly with the Western method. In the instance of a feudal society such as the White Tai n the Tai Highlands, Deo Van

Long ruled his domain so ruthlessly that in many instances he resembled a warlord rather than a president. What is to be reviewed is that use of armed tribal soldiers by indigenous leaders for the purpose of extortion and "protection."

Deo Van Long used his soldiers on numerous occasions for personal political gain, as has already been cited in Chapter III. What is more interesting is also his misuse of his soldiers to subvert a local economy for personal monetary gain. After Long was made the official opium broker of the Tai Highlands by the French Administration, he immediately ordered an increase in opium production. Concurrently, Long lowered the official buying price of opium. Normally only about one-third of the harvested opium is declared by the opium-growing tribes, the rest being used locally and sold to Chinese smugglers for about three times the "official" price. It is through the outside sale of opium that the villagers would receive their outside income, usually in the form of silver.

Deo Van Long and his soldiers would incur the wrath of virtually the entire Hmong population in the Tai Highlands by demanding to "buy" the undeclared opium. Whereas the Hmong had always felt they were being cheated of their opium profits (McAlister, 1967:820), the use of Tai soldiers to force the "buying" of this product at ridiculously low prices now left no doubt (McCoy, 1972:103). The evidence seems to indicate that the outside interference by Tai soldiers of a tribal economy determined the allegiances of the Hmong as a whole at the Battle of Dien Bien Phu (see Chapter III).

During the Second Indochina War, according to Special Forces soldiers interviewed by this author, Montagnard village chiefs and even district chiefs did on occasion resort to demanding personal payment of "protection" from other chiefs who did not possess their own village militia units. The problem of a clandestine crime syndicate forming out of an indigenous tribal army may only appear after the war is over and demobilization is called for. Although specific information on Vietnam is presently lacking, some post-war political developments in the Golden Triangle of Southeast Asia (the tri-border area of Burma-Laos-Thailand) may provide an analogy.

During World War II Japan invaded Burma, the eastern-most possession of the British Empire. The Burmese Army had proven ineffective to meet the overwhelming force of the Japanese invasion. As a result, British Intelligence (SOE) and their OSS American counterparts began arming local tribesmen in the eastern Shan, Kachin, and Karen States. The arming of tribesmen grew apace and eventually over 8,000 Kachin and 12,000 Karens were armed by the Americans and British respectively (Asprey, 1975:608, 611).

As one example, the Shan States had been granted local autonomy by the British since the 1920's, who ruled the states through the indigenous princes, or *sawbwas*. When Burma became independent from England after World War II, the 1947 constitution of Burma promised independence for the Shan States after ten years. By 1958 it was evident to Shan leaders that the Burmese government was going to renege on their agreement and the Shan leaders began rebelling.

When the Shan Rebellion began in January 1958, there were only three or four "armies" in the Shan States, but none were large enough to mount any real threat to the government. The Shan National Army was established in 1961 under the leadership of U Ba Thien and Sao Gnar Kham, a former practicing Buddhist monk. U Ba Thien was no stranger to guerrilla operations. During World War II he organized Lahu tribal guerrilla units for British SOE operations (McCoy, 1972:309). After the war he was asked by the CIA to again form Lahu intelligence units for operations inside communist China. When U Ba Thien later decided to devote his talents to the Shan nationalist cause, he was unable to receive any support in return from his former nation-state employers. As a result, U Ba turned to trading the locally produced opium for modern weapons in much the same manner as Operation X.

This practice in turn led to political bickering within the ranks of the Shan National Army and the eventual alienation of U Ba Thien and Sao Gnar Kham as the principle leaders. According to a journalist:

> " . . . it would be far more accurate to describe the SNA as a grouping o independent warlords loosely tied into a weak federation with a president as figurehead. The president has influence through the facilities he offers for selling opium and buying guns and because he presents a front to the outside world, but it is unlikely he will ever yield effective power unless

> he becomes the channel for outside aid in the forms of guns or money" (McCoy, 1972:313, italics added).

The Shan Rebellion was unable to attract legitimate outside aid, either in the form of exerting pressure on the Burmese for negotiations, or in the form of military assistance once negotiations broke down. The Shan States are rich in minerals, and U Ba Thien even sent samples to the U.S. offering a trade agreement for support in their bid for independence, a legal right in view of the Burmese constitution. As a result of Western disinterest, the Shans were forced to rely on an already existing illicit trade to fund their rebellion.

The Burmese recognized this problem and began forming "self defense" groups in 1963 known as *Ka Kwe Ye* (KKY) for the express purpose of denying the Shan rebels their opium. The government gave the KKY no money and few weapons, expecting them to subvert the opium trade for their own survival. The obvious logic behind this counterinsurgency program was that the Shan Rebellion would weaken once their source of revenue was taken away from them. Instead, this program has added to the number of Shan warlords, since once they become strong enough, many KKY leaders break with the Burmese government and become "independent" (McCoy, 1972:314, 336). This increasing number of Shan rebel leaders and opium warlords, as a direct result of the Burmese counter-insurgency practice, has led to an almost impossible situation in the Shan States:

> " . . . rather than producing in independent, unified Shan land, the Shan rebellion seems to have opened a Pandora's box of chaos that has populated the countryside with petty warlords and impoverished the people. When the rebellion began in 1958, there were only three or four rebel groups active in the entire Shan States. In mid 1971 one Shan leader estimated that there were more than a hundred different armed bands prowling the highlands. But he cautioned that 'it would take a computer to keep track of them all'. Most of these armed groups are extremely unstable; they are constantly switching from rebel to militia (KKY) status and back again, splitting in amoeba like

fashion to form new armies, or entering into ineffectual alliances" (McCoy, 1972:332).

The instance of the Shan States in Burma is not an isolated incident as virtually all of eastern Burma (i.e. Kachin, Wa, Karen states, etc.) is in a permanent state of rebellion. Rebellion in the Kachin State offers a particularly thorny problem for Burma. Although consisting of only 1-3% of the total population of Burma, during World War II the Kachins composed twenty-five percent of Burmese conventional and unconventional troops fighting the Japanese (Johnston, 1966:112). When World War II ended and demobilization was called for, it is obvious that such well-trained, experienced, and unemployed soldiers would be at a premium for any leader with political or criminal ambitions.

This may be the heart of the problem in trying to stabilize a region after prolonged unconventional warfare. Do the U.S. State Department and the U.S. Agency for International Development truly expect former guerrilla fighters to return to their devastated farms as lowly peasants? Do they truly expect militarized tribal groups to accept Barth's first or second option when the third option (total independence) may seem more appealing?

In the instance of Burma, many former guerrillas did not lay down their weapons.[17] In Burma alone there are over a dozen well-organized politico-military factions: the Shan United Army, Shan State Army, Shan United Revolutionary Army, Karen National Union, Kayah New Land Party, Kachin Independence Organization, Pa-O National Organization Army, and the Lahu State Liberation Army to name a few (McCoy, 1972; Coyne, 1984).

Not all of the above-named organizations depend on opium, and even some are violently opposed to any connection to drugs as a source of revenue (Brown, 1984:68). But to maintain an "army" virtually all must resort to either smuggling contraband items (such as teakwood by the Karen National Liberation Army) or be demanding "taxes" at gunpoint from drivers transporting goods through tribal areas. Unfortunately, the Drug Enforcement Agency has only recently discovered that the suppression of one form of contraband may concurrently weaken the position of a legitimate nationalist organization while strengthening the position of an illegitimate rebel organization (Shannon, 1984:63).

If some ethnic groups (or militarized political factions) resort to crime as discussed ion this section, what about the legitimate ethnonationalist organizations themselves? Obviously, some of the above-mentioned tribes in Burma learned how to organize not only legitimate guerrilla organizations, but also illegitimate (criminal) organizations form their nation-state benefactors. Did the Montagnards of South Vietnam also learn how to organize their own militant nationalist organization (FULRO) from their nation-state benefactors?

Ethnonationalist Movements: A Counterinsurgency Byproduct?

Southeast Asia would be wracked by a number of ethnonationalist revolts after the arrival of the French. Some of these revolts by tribesmen were specifically anti-French, while others were anti-monarchical or anti-"lowlander." Virtually all were subdued by military means, often quite brutally, which in turn left smouldering animosities. Some of these anti-French tribal animosities would surface again a generation later and determine the tribal choice for allies during the First Indochina War.[18]

One of the first revolts was led by Pou Kombo, a Kouy tribesman from eastern Cambodia. In an explicitly anti-French revolt he led a sizeable force in 1866 of at least 5,000 which included Khmers, Vietnamese, Chams, Stiengs, and Tagals.[19] In his recruiting drive, Pou Kombo was able to excite peasants and tribesmen alike with the promise that they would all first march to Hue to kill all the French and then on to Udong to do the same (Kiernan, 1982:2). Pou Kombo was defeated and executed in 1867, but a lieutenant of his escaped and returned in 1872 with an army of "400 men of every race in Indochina" (Kiernan, 1982:2).

Similar rebellions would occur in 1876-7 and 1885-6. Both of these rebellions were led by Prince Si Votha of Cambodia. Again in 1916 there would be an "international" revolt involving over 100,000 peasants in Cambodia and South Vietnam. The headquarters for the revolt was also on the Cambodian-South Vietnamese border at Chaudoc (Kiernan, 1982:4).

From 1917-19 another messianic revolt would break out among the Hmong in Laos and North Vietnam. Termed the "Fool's War" by the French, the revolt was under the leadership of Pa Chay, a Hmong shaman:

> "Carrying banners, which Pa Chay assured them would make the invincible, and led by a virgin called Ngoa Nzous, the Hmong rebels were at first successful. Before battle Pa Chay called on ancestors, recited magic formulae and distributed sacred water to his troops. But, it is said, his powers were destroyed when Ngoa Nzoua lost her virginity" (Bessac and Bessac, 1982:69-70).

The first non-religious ehtnonationalist movement, which is to say the first lacking messianic or revivalist overtones, was the "Front pour la Liberation des Montagnards" (FLM). This group began to organize in South Vietnam's Central Highlands in early 1955, barely one year after the French defeat at Dien Bien Phu. The organizer and leader of the FLM was Y Thih Eban, a French-educated bureaucrat who had been raised in the Rhade village of Buon Ale-a.

Y Thih began his education at the Groupe Scolaire Automarchi in Ban Me Thout and the Franco-Jarai School in Pleiku. He received a Primary School Certificate in 1949, after which he went on to the College Sabatier for a clerical training course. Shortly thereafter, in 1950, Y Thih became a Christian and joined the Protestant church in Buon Ale-a. Although Y Thih had many opportunities to contact other Montagnards and become politically active during his education, it seems this was not the case. According to Hickey, Y Thih first became interested in forming a political force in 1955 during which time Y Thih had been an accountant in the Darlac Provincial Health Service, then under French control (Hickey, 1982b).

Being an accountant, Y Thih had noticed an increase of lowland Vietnamese into his own province of Darlac due to the opening up of the land to settlers. Y Thih felt the need to organize a political force in order to persuade President Diem to give the Montagnards the same political and land ownership rights they had enjoyed under the French.

Y Thih began contacting other Montagnards around him in order to see how many others shared his viewpoints toward Vietnamese encroachment. He actually did not have to go much beyond his own immediate environment to find like-minded dissident Montagnards. As a matter of fact, the FLM was primarily made up of other Montagnard bureaucrats like himself (see Table 17). The officers of the FLM were elected as follows:

From the Health Services: Y Mot Nie Kdam, President; Y Say Mlo Doun Du, Treasurer; Y Thih Eban, Secretary; Y Dhua Buon Dap; Paul Doi; Michel Dong. From the Education Services: Y Mo Eban.

In March of 1955, Y Thih wrote a letter, in Rhade, to President Diem listing various acts of maltreatment of the Montagnards by the lowlanders, particularly by the South Vietnamese soldiers. The letter was not answered and it is not known if Diem had ever received it as it was sent by regular post. The FLM took no other actions per se for all of 1956, but in 1957 the Health Services was reorganized, which meant that the FLM was going to break up. This may have actually been an attempt by the government to stop the politicizing of Dalat Province by scattering the FLM leadership (Hickey, 1982b:51).

Whatever the true reason behind the reorganization of the Health Services, the actual outcome was a higher degree of political conscience among the Montagnards. Each FLM leader in effect became a recruiter for a wider range of disaffected Montagnards. Y Thih was sent to Pleiku and there recruited Y Bham Enoul (also from Buon Ale-a), Y San Nie, and Y Ju Eban, all of whom were Rhade. There were also two Jarai recruited by Thih, Siu Sipp and R'mah Lie.

Aside from discontent and action among these civil servants, word spread of the FLM and the idea of such an organization was very well accepted by Montagnard students. In 1958 the Pleiku and Ban Me Thuot groups merged and there was a re-election of the leadership.

Y Bham Enoul, as eldest, became President

Y San Nie, Vice-President

Y Thih Eban, Secretary-General

Siu Sipp, Treasurer

Y Ju Eban, liaison to other province committees

Lt. Y Bhan Kpour, representaive for highland military personnel

Lt. Touprong Ya Ba, Commander of the FLM Etat Major (Military Headquarters)

Officers for each province included:
Kontum: Paul Nur
Pleiku: Siu Plung, Rahlan Yik, Siu Djit, Siu Ja.
Darlac: Y Bih Aleao, Y Du Nie, Y Mo Eban.
Dong Nai: Youneh Yoh, Touneh Phan K'teh

On May, 1958, the various leaders met in Pleiku, where Y Thih suggested a new name for their movement, Bajaraka, which is an an acronym for Bahnar, Jarai, Rhade, and Koho. Y Thih had also made a flag that he proposed be accepted to represent the new movement:

> "The flag featured a red circle dualistically symbolizing the 'red soil of the highlands' and 'the blood shed by the highland people in their struggle for survival' set on a green background representing 'the green highland forests.' At the upper left corner were four white stars for the four highland provinces of Kontum, Pleiku, Darlac, and Haut Donnai (Dong Nai Thuong). The five points on each star represented the five districts in each highland province" (Hickey, 1982b:54).

Y Thih was to claim that 200,000 Montagnards were to become involved in the Bajaraka movement. This would indicate that possibly one-fourth of the Montagnard population was involved with the Bajaraka. Referring to Table 17, the reader will note that approximately one-fourth of the Montagnard leaders belonged to the Bajaraka. The writing of grievance letters, etc., all non-violent activities, would only incense the Saigon government.

Attempts to break up the Montagnard political movements, both of which were non-violent in nature, may have driven the Montagnards to desperate measures. The most *effective* political movement was the militaristic FULRO in view of the fact that Vietnamese officials began meeting with Montagnards after the FULRO Revolt (see Chapter IV). Events seem to indicate that only after the indigenous political groups were militarized were their protests taken seriously. The militarized bloc of tribesmen either contacted or were contacted by the emerging ethnonationalist leadership bloc.

It was the individuals behind FULRO who rightly realized that only their desires for rights *within* the nation-state were being ignored. In their opinion, the only alternative to second-class citizenship would be the establishment of their own state. In relation to Barth's theory, the South Vietnamese and American governments offered the Montagnards Barth's second option, which in reality was no "option" at all. This offer was rejected by the Montagnards who instead opted for a separate state, to be established by armed revolt if necessary. It is this author's opinion that the FULRO movement could have established a separate state were it not for the more pressing interests of the nation-states concerned. Were it not for the strategic role of the Montagnards in the nation-state plan to "stop communism," the Montagnards might have succeeded in establishing a separatist "state" such as the Karens of Burma.

Summary of Findings: Barth and the Unconventional Warfare Formulae

To return to the unconventional warfare formulae, a possible pattern in relation to Barth's theory may have emerged. In RW = (GW + PW), the eventual outcome was assimilation of tribal minorities in to the larger population. The Western approach was COIN = GW with the primary emphasis being the arming of tribal minorities.

The counterinsurgency formula would be valid only up to a certain point in the war. A lack of Western interest in the political needs of tribesmen would lead to a more realistic formula of counterinsurgency in Vietnam of COIN = (GW) + (PW). In the case study of North Vietnam, Deo Van Long provided his own political motivations for the indigenous population. Deo Van Long referred to the historic and legendary control of the White Tai over the Sip Song Chau Tai, and interpreted the Tai Federation as a renaissance of White Tai hegemony over a separate state.

In the instance of South Vietnam and the Montagnards, America proceeded to arm the tribesmen in much the same *laissez faire* manner as the French had. The Montagnards, as the White Tai, in turn proceeded to provide *their own political interpretation* of their role in the war unbeknownst to their nation-state benefactors. The Montagnard and

White Tai particularist political viewpoints (i.e. tribalism), would be at most times at odds with the nation-state's universalist political viewpoint (i.e. stopping communism, imperialism, etc.).

In both case studies of counterinsurgency, the tribal factions attempted to establish and maintain a new state by military means. The political basis of the abortive new states was established along kinship/quasi-kinship lines *a la* Geertz's theory of primordial ties. Deo Van Long received his support primarily from other factions and ethnic groups into which his family had intermarried (see Table 7). As Montagnard nationalism progressed into the Second Indochina War, there was an increase in intertribal marriages (Hickey, 1982b).

This author does not view this as the introduction of new leadership patterns due to the implementation of counter-insurgency doctrine. The White Tai were extremely adept at seeking and concluding outside alliances as shown in Chapter III. The Montagnard leaders also knew of historical alliances being concluded through intermarriage. Certain intertribal clans could intermarry and according to the Rhade poem *Le chant epique de Kdam Yi*, intermarriage was a means to expand political power:

> "The ruling class of chiefs reinforced its position by intermarriage, and each marriage alliance was further guaranteed by the sororate and levirate wherein a deceased partner was replaced by his brother or her sister. The circuit of alliances was in effect a kind of endogamy for the families" (Hickey, 1982b:38).

Although the evidence for a "traditional" pattern of society being maintained during Western counterinsurgency warfare is meager and understandably inconclusive, there does not seem to be any major shift in "traditional" leadership patterns. As reiterated throughout this thesis, the major reason for a lack of leadership change may be due to the simple fact that Western theoreticians never attempted to change the social structure in the same manner as their communist counterparts.

Before any precise statements can be made concerning the effects of unconventional warfare on tribal populations and leadership, further research is obviously necessary. This author has provided throughout this thesis various examples, ancient and modern, of tribal groups in military service to a nation-state as examples for further research. Although the anthropologist would naturally gravitate towards the effects of specific forms of warfare on tribal groups, nation-state utilization of tribal groups presents a highly complex research problem. Social research has yet to seriously consider the effects of unconventional warfare on the nation-state rather than their respective tribal levies.

In the context of the First Indochina War, Jean Larteguy wrote a novelized account in *The Centurions* (1961) describing a French officer corps brutalized by years of counterinsurgency warfare. Michael Grant, a foremost historian on ancient Rome, attributes the dependence on tribal levies for defense as a causal factor to Rome's disintegration:

> "There had long been Germans in the Roman army; and their enrollment had been considerably increased by Diocletian and Constantine. At that time they were mostly *recruited under personal contract, on an individual basis, to serve under Roman officers*; and for the most part they fought well, seeing the empire not as an enemy but as an employer.
>
> "But then in 382 Theodosius I extended the process decisively and transformed it into something more novel. The German 'allies' or federates whom he now enlisted were not merely individual soldiers any longer, but whole tribes, each recruited under its own chieftains, who paid their men in cash and goods received annually for the purpose from the emperor. Once introduced, this new federate participation in the army grew apace. It was widely criticized as an undermining influence, and yet, since other forms of recruitment had failed, it was probably the best remedy available. And yet it proved a failure, and its failure helped to bring the western empire down" (Grant, 1978:440, italics added).

As with Rome and other empires, today's nation-state superpowers have chosen to enlist whole tribal groups under indigenous leadership in unparalleled numbers. As stated in this thesis, such a policy can in turn lead to a myriad of disastrous consequences. Stanley Karnow, in his definitive work *Vietnam: A History*, very aptly termed Vietnam a war nobody won. As a concluding remark, even a cursory review of unconventional warfare practice would seem to indicate a particularly dangerous mode of warfare, especially in regards to tribal militarization.

BIBLIOGRAPHY

ASPREY, Robert B., *War in the Shadows: The Guerrilla in History*, (2 vol.), Doubleday & Company, Garden City, New York, 1975.

BLAUFARB, Douglas S., *The Counter-Insurgency Era: U.S. Doctrine and Performance*, The Free Press, New York, 1977.

BROWN, Robert K., "State Dept. Helicopters: Use and Abuse," in *Soldier of Fortune* (June 1984), Omega Group Limited, Boulder, Colorado, p. 68.

BLOODWORTH, Dennis, *The Messiah and the Mandarins: Mao Tse-tung and the Ironies of Power*, Atheneum, New York, 1982.

CASSIN, Elena, Jean Gottero, and Jean Vercoutter, *Die Altorientalischen Reich I: Vom Palaeolithikum bis zur Mitte des 2. Jahrtausends*, Fischer Taschenbuch Verlag GmbH, Frankfurt am Main, 1965.

Die Altorientalischen Reiche II: Das Ende des 2. Jahrtausends, Fischer Taschenbuch Verlag GmbH, Frankfurt am Main, 1966.

CORSON, William R., *The Betrayal*, Ace Books, Inc., New York, 1968.

COYNE, Jim, "The Last Battle? Burmese Onslaught Crushes Karen Resistance," in *Soldier of Fortune* (June 1984), Omega Group Limited, Boulder, Colorado, pp—65-73.

DOMMEN, Arthur J., *Conflict in Laos: The Politics of Neutralization*, Praeger Publishers, New York, 1971.

DUIKER, William J., *The Communist Road to Power in Vietnam*, Westview Press, New York, 1981.

FAIRBAIRN, Geoffrey, *Revolutionary Guerrilla Warfare: The Countryside Version*, Penguin Books Ltd., Middlesex, England, 1974.

FALL, Bernard B., *Street Without Joy*, The Stackpole Company, Harrisburg, Pa., 1961.

Viet-Nam Witness 1953-66, Frederick Praeger, New York, 1963.

The Two Viet-Nams: A Political and Military Analysis, Frederick Praeger, New York, 1963.

Hell in a Very Small Place: The Siege of Dien Bien Phu, J.B. Lippincott, Philadelphia, 1967.

GEERTZ, Clifford, "The Integrative Revolution: Primordial Sentiments and Civil Politics in the New States," in C. Geertz (ed.), *Old Societies and New States*, Free Press, New York, 1963, pp. 105-57.

GRANT, Michael, *History of Rome*, Charles Scribner's Sons, New York, 1978.

HICKEY, Gerald, *Sons of the Mountains: Ethnohistory of the Vietnamese Central Highlands to 1954*, Yale University Press, New Haven, 1982a.

Free in the Forest: Ethnohistory of the Vietnamese Central Highlands 1954-1976, Yale University Press, New Haven, 1982b.

HOGARD, Jean, "Guerre Revolutionnaire et Pacivication," in *Revue Militaire d'Information*, No. 280 (January) 1957.

HOSMER, Stephen T., *Viet Cong Repression and its Implications for the Future*, Heath Lexington Books, Lexington, Mass, 1970.

JOHNSTON, Howard J., "The Tribal Soldier: A Study of the Manipulation of Ethnic Minorities," in *Naval War College Review*, Naval War College, Newport, Rhode Island, (January) 1967, pp. 98-143.

KARNOW, Stanley, *Vietnam, A History: the First Complete Account of Vietnam at War*, The Viking Press, New York, 1983.

KEEGAN, John, *The Face of Battle*, The Viking Press, New York, 1976.

KELLY, Francis, *U.S. Army Special Forces 1961-71*, Dept. Of the Army, Washington, 1973.

LEE, Susan and David Hurlich, "From Foragers to Fighters: The Militarization of the Namibian San" in Eleanor Leacock (ed.) *The*

Politics and History in Band Societies, Cambridge University Press, New York, 1982, pp. 327-345.

LENG, Shao Chuan and Norman Palmer, *Sun Yat-sen and Communism*, Praeger, New York, 1960.

McALISTER, John T., "Mountain Minorities and the Viet Minh: A Key to the Indochina War" in *Southeast Asian Tribes, Minorities, and Nations*, Princeton University Press, Princeton, 1967.

McCOY, Alfred W., *The Politics of Heroin in Southeast Asia*, Harper & Row, New York, 1972. and Nina S. Adams (eds.), *Laos: War and Revolution*, Harper & Row, New York, 1970.

PARET, Peter, *French Revolutionary Warfare from Indochina to Algeria: The Analysis of a Political and Military Doctrine*, Frederick Praeger, New York, 1964.

RIESSEN, Rene, *Jungle Mission*, Crowell, New York, 1957.

RUBIN, John, *The Barking Deer*, George Braziller, New York, 1974.

SCHROCK, Joann L., et al., *Minority Groups in North Vietnam*, Ethnographic Study Series, DA Pamphlet #550-110, Department of the Army, 1972.

Minority Groups in the Republic of Vietnam, Ethnographic Study Series, DA Pamphlet #550-105, Department of the Army, 1966.

STERLING, Claire, *The Terror Network*, Berkley Books, New York, 1981.

SHANNON, Elaine, "The Asian Connection" in *Newsweek*, June 25, 1984, Newsweek Inc., New York, pp. 62-3.

TRINQUIER, Roger, *Modern Warfare: A French View of Counter-Insurgency*, Frederick Praeger, New York, 1964.

WEST, F. J. Jr., *The Village*, Harper & Row, New York, 1972.

ZASLOFF, Joseph J., *The Pathet Lao: Leadership and Organization*, D.C. Health & Co., Lexington, Mass, 1973.

ENDNOTES

[1] A conversion to Communism was successful in some Asian Confucian countries, i.e., Vietnam, China, Korea (Blaufarb, 1977:100)

[2] Ho obviously did not believe this himself. In his last testament before his death he stated he was going to join other revolutionary elders such as Karl Marx and Lenin. Ho did not mention Confucius' name, nor any Vietnamese heroes either (Fairbairn, 1974:249)

[3] The foundation for a parallel politico-military hierarchy can be dated back to January 1866 in Russia. Nicholas Ishutin, a Moscow revolutionary, organized and integrated a political group for propaganda and a military group devoted to terrorism. The first group he called "The Organization," the latter "Hell." Lenin adopted this principle for the Bolshevik Revolution (Sterling, 1981:188).

[4] The French recognized five phases (*phases*) of what they termed the rotting away (*pourrissement*). As above it began with agitation-propaganda teams and ended with the general counter-offensive (Hogard, 1957:11).

[5] It will always remain as speculation what would have become of Trinquier and General Giap had there been no war. Giap earned a Ph.D. in History before becoming a soldier, and might have begun a career teaching at the university level had there been no war. Comparatively, Trinquier's only ambition before the army was to become a teacher.

[6] Groupment Commando de Mixtes Aeroportes, translated as Mixed Airborne Commando Group.

[7] Captain Savani was in charge of the SDECE in Indochina. Although this intelligence agency was dominated by the French military, its American equivalent would be the civilian Central Intelligence Agency.

[8] See Corson (1968, 296:309) for a complete listing of the projects and objectives of the RD Program.

[9] The Taiping and Boxer rebellions in China would offer interesting comparative examples to this thesis. The *T'ai P'ing T'ien-kuo* (Heavenly Kingdom of the Great Peace) was a bizarre pseudo-Christian revolt (1847-66) against the Confucianist-Taoist Manchu elite. Espousing a doctrine labeled as "primitive communism," the peasant leader Hung Hsiu-chuan attracted a conglomerate of peasants, intellectuals, coolies, pirates and businessmen. The Boxer Rebellion (1898-1900) was led by the *I Ho Ch'uan* Society (Righteous Harmonious Fists). Both the Boxer movement and the *I Ho Ch'uan* were an outgrowth of provincial militias raised by the Manchus to fight the Taiping's (Asprey, 1975:327,335).

[10] President Diem also refused to negotiate with Ho for similar reasons. The two leaders met once and Ho was asked why he had Diem's brother killed. Ho's answer that mistakes were unavoidable during revolutions did not please Diem (Karnow, 1982:216).

[11] Due to the nature of the GCMA's as explained in Chapter II, the number of tribesmen trained in the Tai Highlands will never be known. One estimate would be 10,000 tribal soldiers in the Tai Highlands. This would mean the pro-French tribal soldiers saw one-fifth of their "army" killed in action in one week (McAlister, 1967).

[12] According to Dr. Haas of the Freie Universitaet Berlin (1980, personal communication) the first documented guerrilla war was fought between the Hittites and the Kaska tribesmen in Anatolia (Turkey) circa 1500 BC. What is interesting about this is that it took place in the mountainous homelands of the Kaskaer with the Hittites at first using costly conventional military tactics.

About 200 years later, circa 1300 BC, the Kaskaer would paly a different role in a monumental battle that is not dissimilar in many aspects to Dien Bien Phu. Both battles were fought by large armies (considering the logistics), both were spectacular, and both were considered decisive.

One glaring fact of overlooked, as with the Battle of Dien Bien Phu, is the role of the tribal levies in the Hittite and Egyptian armies at the Battle of Kadesch. Htusilli III, the brother of the Hittite commander Muwatili, claimed that his personal levies of Kaskaer were decisive at the battle. Egyptian texts mention numerous tribal contingents that also partook, such as the Arzawa, Masa, Lukka, Kizzuwatna, Karkemis and assorted levies from Syria (Cassin, 1966b:156).

[13] The statue of Alexandre de Rhodes in Saigon was torn down by the communists after the defeat of South Vietnam in 1975.

[14] Comparative examples of Japanese-indigenous units abound in Asia. One such is the "JIFs," or Japanese Indian Forces that were formerly part of the British Army.

[15] The Nungs are a rather interesting example of militarization in Southeast Asia. Some sided with the French during the First Indochina War, and after the French defeat 15,000 Nungs arrived in South Vietnam. Most of them had been part of a French Army division (Hickey, 1982b:16).

[16] Self-criticism and the Communist Chinese and Vietnamese obsession with "self-criticism sessions" was adopted from Confucianist principles. Obviously this would be an alien concept to those who were never exposed to Confucianism (Bloodworth, 1982:187).

[17] This is not an isolated instance anywhere around the globe. When the unconventional war in Portugal's African colonies "ended," pro-West UNITA and RENAMO refused to lay down their weapons and are still fighting in Marxist Angola and Mozambique respectively.

[18] For an interesting analysis of second-generation personal animosities determining "allies" in Laos during the Second Indochina War, see *Laos: War and Revolution* (McCoy, 1970).

[19] The Tagals were recruited from the Philippines as palace guards (Kiernan, 1982:2).

TABLE 1

The Uy Ban Hanh Chinh (UBHC)

Village Committee, or UBHC-xa (five members)

1) President

2) Vice-president: in charge of police (usually of former veteran who would also be incharge of village self-defense)

3) Secretary: in charge of paperwork

4) Member: in charge of finances

5) Member: in charge of public works and agriculture

Inter-village Committee, or Lien-xa

More sophisticated than the UBHC-xa, but somewhat similar. An element of military command may start at this level.

District Committee, or UBKC/HC-Huyen

This new designation was now "Committees for Resistance and Administration" (Uy Ban Khang Chien/ Hanh Chinh). This was actually a military organization. In may 31, 1958 the various zones were "demilitarized" which changed them all simply to UBHC's.

Province Committee, or UBKC/HC-tinh

Zone Committee, or UBKC/HC-khu

TABLE: 2

French Intelligence and Counterinsurgency Organizations During the First Indochina War.

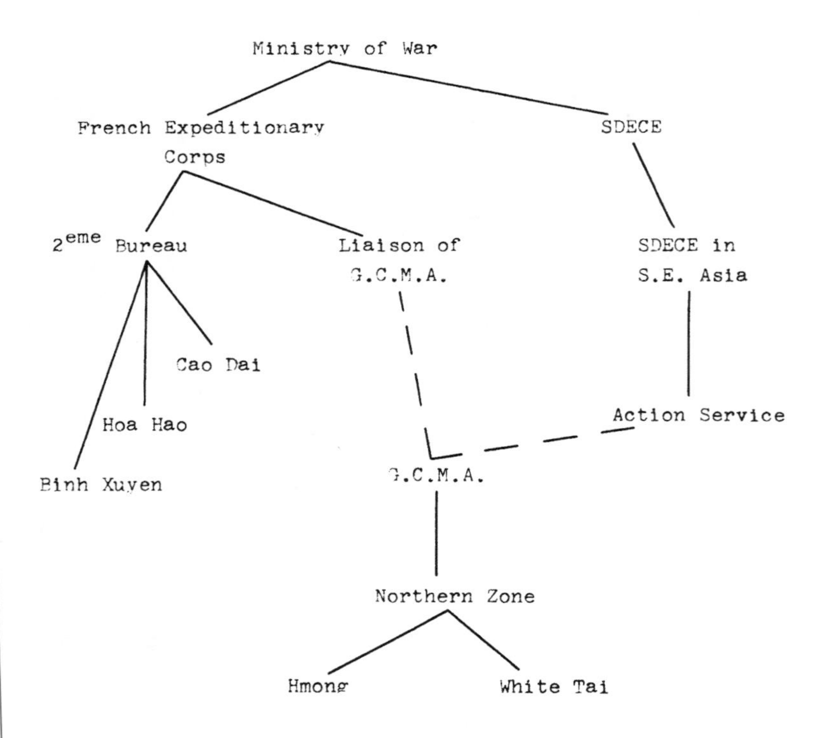

Source: McCoy 1972

TABLE 3

Montagnard Aid Programs From 1964-1970

1) 49,902 economic aid projects
2) 34,334 educational projects
3) 35,468 welfare projects
4) 10,000 medical projects
5) 129 churches built
6) 272 markets established
7) 110 hospitals built
8) 479,568 refugees supported
9) 14,934 transportation facilities
10) 6,436 wells dug
11) 1,003 classrooms built
12) 670 bridges built
13) 398 dispensaries built

(Source; Kelly, 1972:155)

TABLE: 4

Ethnic Composition and Percentages of French Union Forces at Dien Bien Phu on March 1954.

Ethnic Composition	Total	Percent
French Mainland	1,412	13.05
Foreign Legion	2,969	27.45
North Africans	2,607	24.10
Africans	247	2.28
Vietnamese	1,104	10.20
Tribal Tai	2,575	23.82
Total	10,814	100.00

(Source: Fall, 1967)

TABLE: 5

Indigenous and Colonial Units at Dien Bien Phu on March 1954

2/1 Algerian Rifles

3/3 Algerian Rifles

5/7 Algerian Rifles

1/4 Moroccan

2nd Tai Battalion

3rd Tai Battalion

Tai Partisan Mobile Group 1
(11 Companies)

2nd Group, 4th Colonial Artillery

3rd Group 10th Colonial Artillery

11th Battery, 4th Colonial Artillery

8th Commando Group (GCMA)

(Source: Fall, 1967)

TABLE: 6

North Vietnamese People's Army Units at Dien Bien Phu on March 1954

304th Infantry Division (primarily lowland Vietnamese)

- 57th Infantry Regiment
- 345th Artillery Regiment

308th Infantry Division (primarily lowland Vietnamese)

- 36th Infantry Regiment
- 88th Infantry Regiment
- 102nd Infantry Regiment

312th Infantry Division (primarily tribal Tho)

- 141st Infantry Regiment
- 165th Infantry Regiment
- 209th Infantry Regiment
- 154th Artillery Battalion

316th Infantry Division (primarily tribal Tho)

- 98th Infantry Regiment
- 174th Infantry Regiment
- 176th Infantry Regiment
- 812th Heavy Weapons Company

(Source: Fall, 1967)

TABLE: 7

Social Matrix of Tribal Interaction in the Tai Highlands

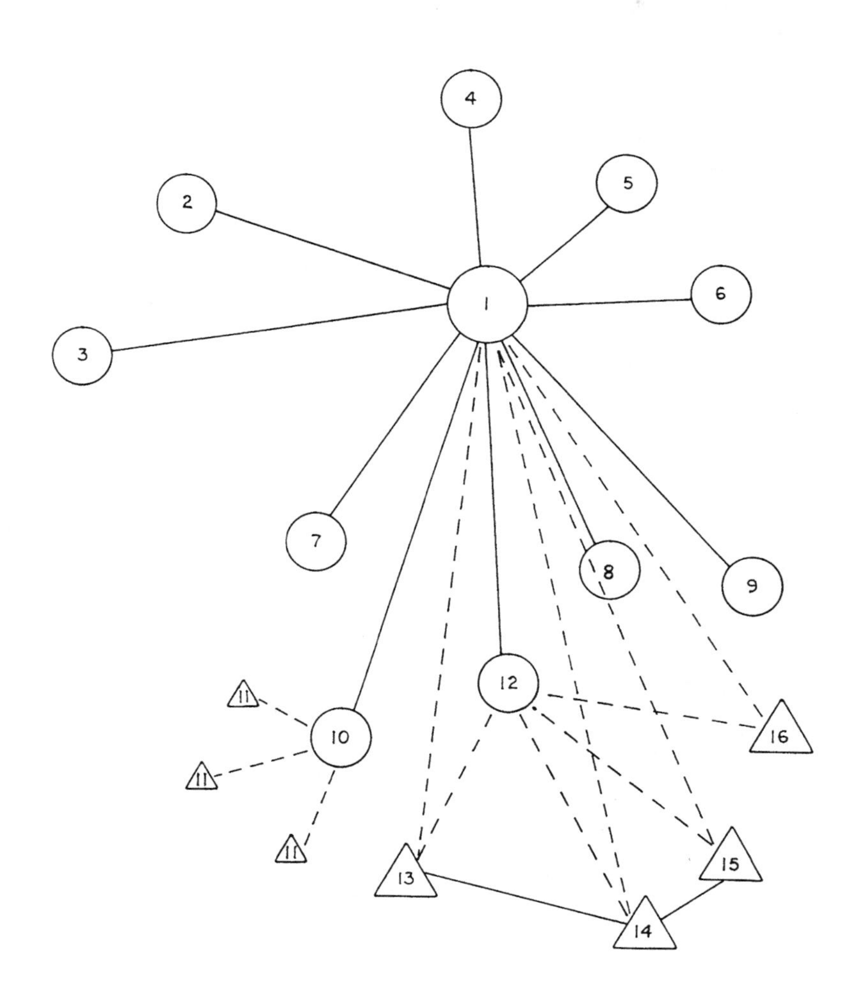

TABLE: 8

Explanatory Table to Diagram.

1. Lai Chau Deo political faction (Tai Highlands)
2. Phong Saly political faction (Laos)
3. Luang Prabang, Laotian Royal Family
4. Pa Tan political faction, Tai Highlands (intermarried with Lai Chau Deo
5. Chieng Chan, Tai Highlands (intermarried with Lai Chau Deo
6. Phong Tho Deo political faction
7. Quinh Nhai, Tai Highlands
8. Phu Yen, Tai Highlands
9. Tuan Giao, Tai Highlands
10. Dien Bien Phu political faction (Deo Van Long replaced Lo Van Hac with his own son)
11. Hmong from the mountains surrounding Dien Bien Phu
12. Thuan Chau political faction, Tai Highlands (Bac Cam Qui)
13. Moc Chau political faction, Tai Highlands (Xa family)
14. Yen Chau political faction, Tai Highlands (Hoang family)
15. Mai Son political faction, Tai Highlands (Cam family)
16. Muong La political faction, Tai Highlands (Cam family)

--- Solid lines indicate mutual support

- - Broken lines indicate mutual antagonism

TABLE: 9

Montagnard Population Estimates for South Vietnam

Rhade	100,000
Jarai	150,000
Bahnar	75,000
Mnong	15,800
Stieng	30,000
Sedang	40,000
Bru	40,000
Hre	100,000

Source: Hickey, 1982:302

TABLE: 10

Equipment and Organizational Table for C.I.D.G. Light Guerrilla Company (150 Men)

Two Company Headquarters Elements, 10 enlisted men each

Three Rifle Platoons, 35 enlisted men each

Three Weapons Platoons, 35 enlisted men each

Equipment	Quantity
Carbine, .30-calibre, M1	103
Launcher, grenade carbine	24
Machine gun, light, .30-calibre	3
Mortar, 60mm	3
Pistol, .45-calibre	1
Radio, HT-1	20
Radio, TR-20	2
Rifle, automatic, Browning	18
Submachine gun, .45-calibre, M3	29

Source: Kelly 1972

TABLE: 11

Ethnic Composition of Strike Forces

Campsite	Strike Force Co.	Men	Nungs	Ethnic Composition of Strike Force	Remarks on Area
Darlac Province					
Buon Mi Ga	4	619	0	Mnong	Mnong and Rhade tribal groups in area
Ban Don	5	799	0	Rhade, Mnong, Jarai, Thai	Rhade and Jarai tribal groups in area. Part of Buon Enao
Buon Brieng	3	775	22	Rhade	Rhade tribal group in area
Pleiku Province					
Pleiku	0	0	60	Nung, Rhade, Vietnamese	Jarai tribal group in area
Plei Do Lim	2	271	41	Jarai, Bahnar, Nung	Jarai tribal group in area
Plei Djereng	2	231	30	Jarai, Bahnar	Jarai and Rhade tribal groups in area
Duc Co (Chu Dron)	4	444	35	Jarai, Nung	Bahnar and Jarai tribal groups in area
Plei Me	3	241	35	Jarai, Nung	Jarai tribal group in area
Plei Mrong	3	464	37	Jarai, Bahnar, Rhade	Jarai tribal group in area

TABLE: 31 (continued)

Campsite	Strike Force Co.	Men	Nungs	Ethnic Composition of Strike Force	Remarks on Area
Kontum Province					
Dak Pek	4	479	33	Sedang, Jarai, Halang Vietnamese, Nung	Sedang tribal group in area
Polei Krong	4	463	25	Rongao, Jarai, Bahnar	Bahnar, Jarai, Sedang, and Mnong tribal groups in area
Dak To	3	371	0	Sedang, Nongao, Bahnar	Sedang tribal group in area
Khanh Hoa Province					
Dong Ba Thin	2	308	9	Vietnamese, Cham, Tuong, Mien	
Nha Trang	3	505	0	Vietnamese, Cham, Rhade, Raglai, Nung	Cham and Rhade tribal groups in area
Binh Dinh Province					
Kannack	3	418	36	Bahnar, Rhade, Jarai, Bong, Mien, Nung	Rhe tribal group in area
Plei Ta Nangle	4	606	0	Bahnar, Jarai, Nung	Bahnar and Rhade tribal groups in area
Quang Duc Province					
Bu Prang		no figures		Rhade, Mnong	Mnong tribal group in area

TABLE: 11 (continued)

Campsite	Strike Force Co.	Men	Nungs	Ethnic Composition of Strike Force	Remarks
Phu Yen Province					
Dong Tre	4	619	51	Bahnar, Nung	
Tuyen Duc Province					
Phey Srunh	4	652	0	Koho, Ma, Chil	Koho tribal group in area
Phu Bon Province					
Buon Beng (Cheo Reo)	5	865	0	Jarai, Bahnar, Drung	Bahnar and Jarai tribal groups in area
Quang Tri Province					
Khe Sanh	5	677	48	Vietnamese, Bru	Bru tribal group and lowland Vietnamese in area
Thua Thien Province					
A Shau	3	332	45	Vietnamese, Tau-Oi	Lowland Vietnamese in area
Phuoc Long Province					
Bu Ghia Map	4	449	9	Vietnamese, Cambodian, Stieng	Stieng tribal, Vietnamese and Cambodians in area
Bu Dop	3	275	0	Stieng	Stieng tribal group in area

TABLE: 12

The Tho Political System From Village to Province

Hamlet (ban)	po ban: chief cai thon
Village (xa)	ly truong/xa truong: chief
Canton (tong)	chan tong: chief pho tong: secretary tong doan: defense group
Chau	quan chau: chief lai muc: assistant thong lai: assistant - - : secretaries chau doan: head of guard forces linh: subsidiary guard doi le: leader of subsidiary guard
Province (tinh)	tuan phu/tong doc: chief bo chanh: administrative assistant an sat: judicial chief doc hoc/kiem hoc: educational overseer de doc/lanh binh: military leader

TABLE: 13

Correlation of Traditional Leadership Positions to Communist Parallel Hierarchies

Traditional	Tho	Meo	White Tai	Black Tai	UBHC/UBKC
I. Political					
1. Hamlet chief	po ban				
a. assistant					
2. Village chief	ly truong/ xa truong	seo phai seo phao	ly truong	ly truong	Pres. UBHC-xa
a. assistant					Secretary
3. Canton chief	quan chau	ping tau	chau muong	chau muong	Pres. UBKC/HC-Huyen
a. assistant	lai muc		(tho lai)	tho lai	Secretary
4. Province chief	tuan phu/ tong doc	sung quan			Pres. UBKC/HC-tinh
a. assistant	bo chanh				Secretary
II. Military					
1. Hamlet	po ban?				
a. assistant					
2. Village	ly truong	seo phai			Vice-Pres. UBHC-xa
a. assistant					

TABLE: 13 (cont'd)

Traditional	Tho	Meo	White Tai	Black Tai	UBHC/UBKC
3. Canton	chau doan	ping tau	chau muong	chau muong	Vice-Pres. UBKC/HC-Huyen
a. assistant	doi le				
4. Province	de doc/ lanh binh				Vice-Pres. UBKC/HC-tinh
III. Kinship					
1. Aristocratic clans	Tho-ti	none	Lo, Cam	Lo, Cam	Aristocrats not recognized
2. Commoner clans			Vi, Kuan, Ka Lu, Non, Dan	Vi, Lu, Leo Tong, Kwang Ma, Nguyen	All considered equal under Communist system
IV. Land Ownership	males (private)	household (private)	chau muong	chau muong	Private land ownership abolished

TABLE: 14

Simplified Social Matrix of Village Social Organization Before Imposition of Communist Front Organizations

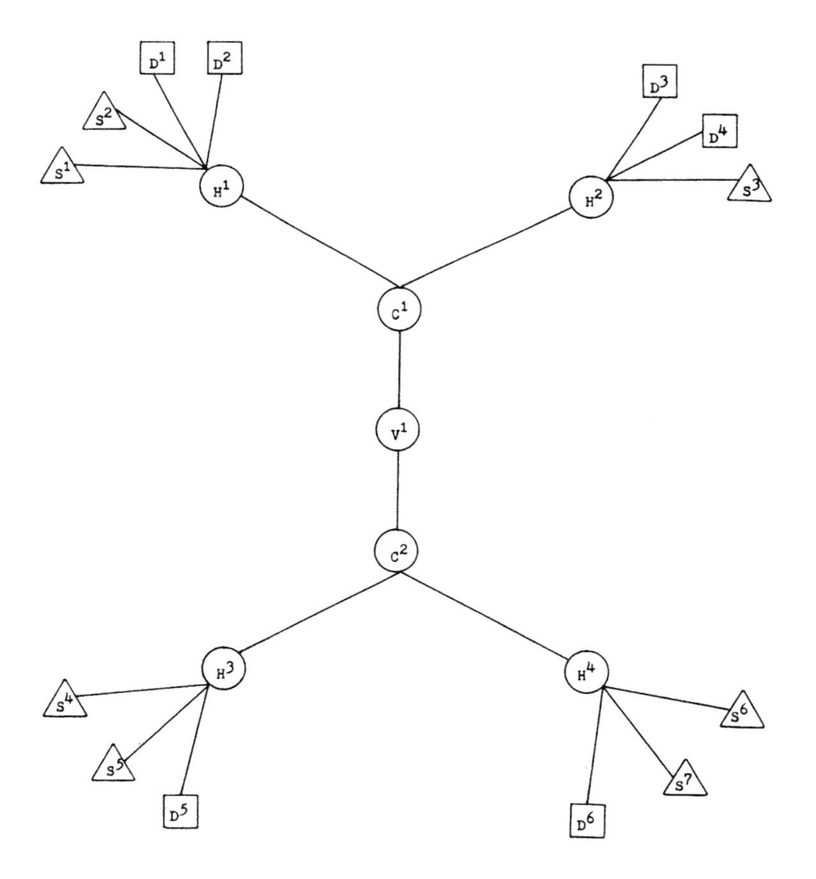

TABLE: 15

Simplified Social Matrix of Village Social Organization After Imposition of Communist Front Organizations

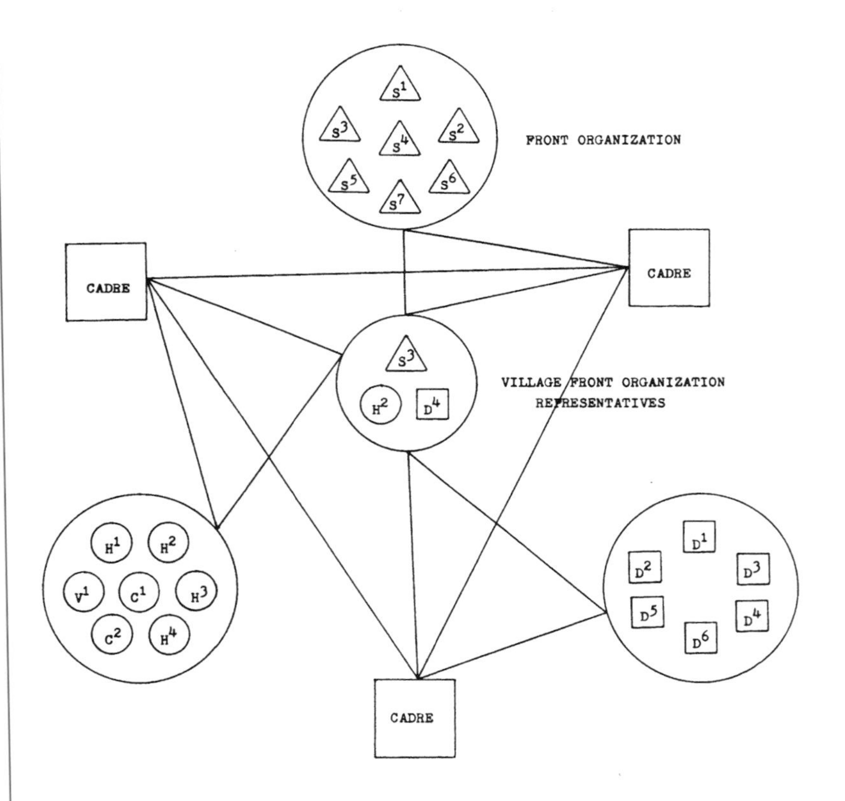

TABLE: 16

Explanatory Table to Social Matrix 2 and 3.

= son

D = daughter

H = male head of household

C = male head of clan

V = male village leader

For Table 14 solid lines indicate individual hierarchical lines of communication to village chief

For Table 15 solid lines indicate group hierarchical lines of communication to communist cadre and village representatives

TABLE: 17

Individual Remarks Pertaining to Gerald Hickey's Leadership Group

Name/tribe	DOB	REL	MS	BS	FL	BA	FU	CS	LY	Remarks
Jarai										
Kpa Koi	1934	T					X			Headed seperate branch of FULRO
Kpa Doh	1937	T	X			X	X			Interpreter at SF camp Bon Beng. Colonel by 1968 in FULRO. Went to Mondulkiri
Ksor Wol	1908	T								Assistant to King of Fire
Ksor Glai	1917	C								Judge in Pleiku Highlander court
Ksor Dhuat	1937	T				X	X			1964 Revolt organizer. Went to Mondulkiri.
Ksor Dun	1933	P						X	X	Liaison between FULRO and RVN in 1964 Revolt
Ksor Rot	1937	T	X				X			Elected to National Assembly
Ksor Hip	1942	T								
Ksor Kok	1944	P					X			Went to Mondulkiri
Nay [illegible]er	1900	T								Graduate Southern Ethnic Minorities School, Gia Lam (N. Vietnam)
Nay Moul	1913	T	X							One of first Jarai to join French Army, 2nd Lt. Served in 5th Batt., FIW

TABLE: 17 (continued)

Name/tribe	DOB	REL	MS	DS	FL	BA	FU	CS	LY	Remarks
Nay Phin	1916	T								Graduate Southern Ethnic Minorities School, Gia Lam (N. Vietnam)
Nay Lo	1919	T	X							Major in ARVN. Became Pleiku province chief
Nay Dai	1925	T	X							Served in 5th Batt., FIW
Nay Blim	1928	T						X		National Assembly. Lower House
Nay Honh	1933	T	X							Served in 5th Batt., FIW. Capt. in ARVN, SIW
Nay Luett	1935	C		X		X		X	X	Served in MDEM
Nay Alep	1940	C		X						Graduate National Inst. of Administration
Rcom Rock	1920	B		X		X				
Rcom Briu	1922	T	X	X						In 1961 Secretary, Central Highland Autonomy Movement. Later General in NVA
Rcom Pioi	1925	T	X					X		Teacher. 1st Lt. in Army. Served in 5th Batt., FIW. Brother became General in NVA
Rcom Hin	1927	T	X			X				Former school teacher. Served in French Army 1942, 5th Batt. FIW, Lt. Colonel in ARVN, SIW
Rcom Perr	1934	C				X		X		

TABLE: 17 (continued)

Name/tribe	DOB	REL	MS	BS	FL	BA	FU	CS	LY	Remarks
Rcom Anhot	1936	T						X	X	Graduate National Inst. of Administration. National Assembly
Rcom Po	1937	T								Elected to Ethnic Minorities Council
Rcom H'ueh	1941	T								
Rcom Nhut (Ali)	1947	T								Assistant to CORDS rep. Ed Sprague
Rahlan Beo	1922	T								
R'mah Liu	1934	T		X	X			X		
R'mah Wih	1942	T	X							Capt. in ARVN, SIW
Ro-o, Bleo	1943	T								
Siu Plung	1926	T	X		X	X				Graduate Gendarmerie Nationale
Siu Sipp	1929	T	X	X	X	X				Sergeant in French Army. Killed by Viet Cong, 1963
Siu Nay	1934	T	X					X		Served in 5th Batt., FIW
Siu Ky	1943	C								
Rhade										
Adrong, Y Dhe	1920	P				X	X			

TABLE: 17 (continued)

Name/tribe	DOB	REL	MS	BS	FL	BA	FU	CS	LY	Remarks
Adrong, Y Dhon	1933	C		X		X	X		X	Director of primary education at Lac Thien. 1964 Revolt organizer. Went to Mondulkiri. Executed by FULRO Dec. 1965
Adrong, Y Klong	1934	P						X		Interpreter for Village Defense Program
Aleo, Y Bih	1901	T	X	X	X	X				Served in Guarde Indochinoise. Head of Darlac FLM and Bajaraka. Village headman. 1961 Chairman, Central Highlands Autonomy Movement (Viet Cong)
Buon Dap, Y Dhua	1923	T		X	X	X				Founding member of Bajaraka
Buon Krong, Y Preh	1920	P		?	X	X	X			Affiliated with Christian and Missionary Alliance. Interpreter at Buon Enao. Colonel in FULRO
Buon Krong Pang, Y Bling	1922	C		X			X			Former teacher. FULRO Colonel
Buon To, Y Jut	1944	C						X		Rhade leader
Buon Ya, Y Puk	1940	T								
Buon Ya, Y Wik	1941	T					X			Vice-Minister of Health, FULRO. Later served in National Assembly. Executed by Communists during Tet 1968
Buon Ya, Y Ngo	1943	C					X			Executed by Communists during Tet 1968
Eban, Y Sok	1900	T	X							Garde Indigene 1921. One of first recruits from Franco-Rhade School

TABLE: 17 (continued)

Name/tribe	DOB	REL	MS	BS	FL	BA	FU	CS	LY	Remarks
Eban, Y Bloc	1917	T	X							Commanded 120th Viet Minh Regiment in 1954
Eban, Y Mo	1927	T		X	X	X		X		
Eban, Y Thih	1932	P		X	X	X	X	X		Founder of Bajaraka. Repelled by FULRO violence. Graduate of Franco-Jarai School
Eban, Y Ju	1935	T		X	X	X	X	X		Founding member of Bajaraka. Went to Mondulkiri
Eban, Y Nham	1937	T	X							Lt. in ARVN
Eban, Y Du	1949	T								
Enoul, Y Bham	1913	P		X	X	X	X			Graduate Ecole Nationale d'Agriculture. FULRO leader. House arrest in Cambodia Dec. 31, 1968
Hmok, Y Blieng	1920	T		X						
Hwing, Y Tin	1944	T								
Kbuor, Y Soay	1917	T								
Knuol, Y Pem	1927	T	X			X				Joined French Army in 1943
Kpuor, Y Bhan	1937	T	X		X	X	X			Lt. in ARVN. Went to Mondulkiri
Mlo, Y Kdruin (Philippe Drouin)	1936	C	X			X	X			Interpreter at SF camp. Served in Cambodia in French Army. Colonel in Dam Y Division of FULRO

TABLE:17 (continued)

Name/tribe	DOB	REL	MS	BS	FL	BA	FU	CS	LY	Remarks
Mlo Duon Du, Y Toeh	1910	P								School teacher
Mlo Duon Du, Y Wang	1925	T								
Mlo Duon Du, Y Say	1925	T		X	X	X				Founding member of Bajaraka. Killed during Tet 1968
Mlo Duon Du, Y Chon	1933	P		X		X		X		Served in MDEM. Was pro-FULRO
Nie Buon Drieng, Y Blu	1923	T		X						Liason between FULRO and RVN during FULRO Revolt 1964
Nie Hrah, Y Ham	1930	P								
Nie Kdam, Y Dhuat	1919	T		X		X				Liason between FULRO and RVN during FULRO Revolt 1964
Nie Kdam, Y Ngong	1923	T		X						Joined Viet Minh in 1945. One of few highland leaders trained in Russia
Nie Kdam, Y Sen	1943	C	X			X	X			Army officer in ARVN security forces. Left army to join FULRO. Became FULRO Major. May have been gov't spy
Chru										
Banahria Ya Don	1917	T		X						Joined Viet Minh in 1945. In interrum worked in administration. Joined Viet Cong in 1961
K'Kre	1919	C		X						Member of Dalat City Council

TABLE: 17 (continued)

Name/tribe	DOB	REL	MS	BS	PL	BA	FU	CS	LY	Remarks
Touneh Han Din	1917	T		X						Highlander Law Court 1966. Killed in helicopter crash 1966
Touneh Han Tin	1922		X			X				7th Batt. officer FIW. Capt. in ARVN while training Mountain Scout soldiers SIW. Died of fever 1964
Touneh Yoh	1936	P		X	X	X		X	X	
Touneh Ton	1936	C	X					X		7th Batt. sergeant FIW
Touprong Hiou	1917	T		X						Arrested as Viet Cong 1955-61
Touprong Ya Ba	1923	P	X		X	X				Joined French Army 1941. Commander of 7th Batt. FIW. Became Lt. Col. after averting violence in FULRO Revolt 1964
Ya Yu Sahau	1923	C		X		X				School teacher
Ya Duck	1933	C					X			FULRO Major
Bahnar										
Doi, Paul	1935	C		X	X	X				Founding member of Bajaraka
Dong, Michel	1935	C		X	X	X				Founding member of Bajaraka
Hiar	1905	C		X						One of four Highlanders in Diem's gov't. School teacher in Kontum
Hiu	1930	C								

TABLE: 18

Ethnic Composition and Percentage of Gerald Hickey's Leadership Group

Tribe	Population	% of Montagnard Population	% of Hickey's Leadership Group	Ratio of Leaders to Tribal Population
Jarai	150,000	24.0%	35%	1:4,285
Rhade	100,000	16.0%	34%	1:2,941
Chru	15,000	2.4%	10%	1:1,500
Bahnar	75,000	12.0%	7%	1:10,714
Mnong-Rlam	40,000	6.4%	3%	1:13,333
Bru	40,000	6.4%	2%	1:20,000
Hre	100,000	16.0%	2%	1:50,000
Lat	1,300	.2%	2%	1:600
Sedang	40,000	6.4%	2%	1:20,000
Sre	22,000	3.5%	2%	1:11,000
Katu	40,000	6.4%	1%	1:40,000

TABLE: 19

Montagnard Tribes From First Nation-State Militarization to 1964 FULRO Revolt

Tribe	1	2	3	4	5	6	7	8	9	10	11	12	13
Bahnar	X		X	X	X	X	12%	7%	0%	71.4%	43%	57%	0%
Bru						X	6.4%	2%					
Chil					X	X							
Chru							2.4%	10%	30%	60%	20%	40%	10%
Halang						X							
Hre				X		X	16%	2%					
Jarai	X		X	X	X	X	24%	35%	37%	17%	8.5%	23%	14%
Katu							6.4%	1%					
Koho						X							
Lat							.2%	2%					
Maa				X		X							
Mnong				X		X	6.4%	3%					
Nongao						X							
Raglai						X							
Rengao	X					X							
Rhade		X	X	X	X	X	16%	34%	23%	41%	25%	44%	38%
Sedang	X		X	X	X	X	6.4%	2%					
Sre							3.5%	2%					
Stieng						X							

TABLE: 20

Explanation to Table

Column 1: Recruited for Mayrena's Bahnar-Rengao Confederation and the Kingdom of Sedang

Column 2: Recruited for Sabatier's Garde Indigene, circa 1930

Column 3: Recruited during World War II for the 4eme Bataillon Montagnard du Sud Annam

Column 4: Recruited during First Indochina War for the Division Montagnard

Column 5: Recruited during First Indochina War for Mobile Groups 41, 42, and 100

Column 6: Recruited during Second Indochina War for Strike Force (refer to Table)

Column 7: Percent of Montagnard population

Column 8: Percent of Hickey's list of 100 Montagnard leaders

Column 9: Percentage per Hickey's list with military background

Column 10: Percentage per Hickey's list with background as bureaucrats

Column 11: Percentage per Hickey's list who were involved in the F.L.M.

Column 12: Percentage per Hickey's list who were involved in the Bajaraka

Column 13: Percentage per Hickey's list who were involved in FULRO